Reverse Engineering

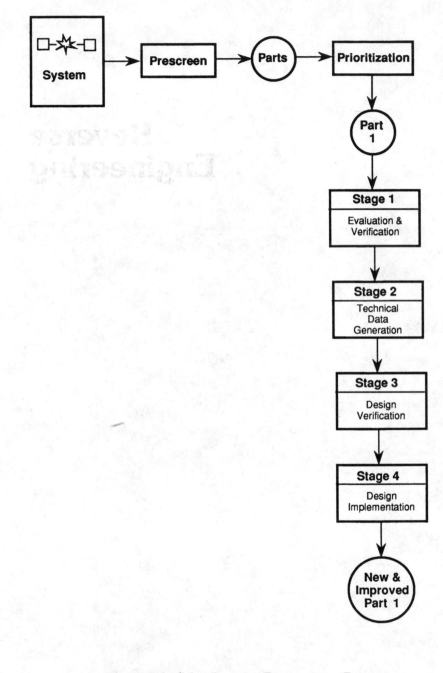

Overview of the Reverse Engineering Process

Reverse Engineering

Kathryn A. Ingle

McGraw-Hill, Inc.

New York   San Francisco   Washington, D.C.   Auckland   Bogotá
Caracas   Lisbon   London   Madrid   Mexico City   Milan
Montreal   New Delhi   San Juan   Singapore
Sydney   Tokyo   Toronto

Library of Congress Cataloging-in-Publication Data

Ingle, Kathryn A.
   Reverse engineering / Kathryn A. Ingle.
     p.   cm.
   Includes bibliographical references and index.
   ISBN 0-07-031693-7
   1. Reverse engineering. I. Title.
   TA168.5.I64   1994
   658.5—dc20   94-18447
                                                       CIP

2 3 4 5 6 7 8 9 0 DOC/DOC 9 0 9 8 7 6 5

ISBN 0-07-031693-7

The sponsoring editor for this book was Robert W. Hauserman, the editing supervisor was Caroline Levine, and the production supervisor was Donald F. Schmidt. It was set in Baskerville by Carol Woolverton Studio, Lexington, Massachusetts, in association with Warren Publishing Services, Biddeford, Maine.

Printed and bound by R. R. Donnelley & Sons Company.

Contents

5. Stage 2: Technical Data Generation 85

6. Stage 3: Design Verification 107

7. Stage 4: Project Implementation 117

8. Summary of Reverse Engineering 129

9. Future Applications 139

Preface

I do not believe that reverse engineering is truly anything new although its origins are vague. In all the research I have done on the subject since 1985 I have found no definitive discussion of where it came from or how it is conducted. In late 1984–early 1985 there was a single line in the U.S. Government Federal Acquisition Regulation that stated reverse engineering was to be considered only when it was economically feasible and all other options had been exhausted as a way to develop technical data. At that time the objective was to acquire competitively as many hardware parts as possible to combat the embarrassing $400 hammers that had gotten the attention of the public and which, of course, riled Congress. The drive to develop technical data fueled the need for reverse engineering.

Of the thousands of mechanical and electrical parts to review for potential acquisition through competitive procurement, relatively few were considered candidates for reverse engineering at first. The relative few became a few hundred—and later a few thousand. This potential pool of projects drove the need for a reliable method which could determine where to invest the economic resources available for this task. The program as originally envisioned had to provide three crucial components: (1) a reliable screening procedure, (2) accountability of funds, and (3) a return on investment. What follows is a distillation of what was learned in 5 years.

Over 2000 candidates were screened.

Approximately 200 parts passed the screening and were evaluated, and about 150 projects were inducted.

The overall return on investment was 23.8 to 1 for a program budget of over $2 million.

Since the inception of this limited effort I have found that 90–95 percent of the literature in circulation today about reverse engineering is aimed toward software reverse engineering with the remaining 5 or 10 percent aimed at hardware systems. Much of this 5 to 10 percent is directed at pieces of the process, such as dimensional inspection techniques or digitizing the data. At most, only stages 2 and 3, technical data generation and design verification, are discussed. Few, if any, documents give guidelines for the selection of candidates, data collection, or the economic justification for reverse engineering. The purpose of this book, then, is to pull all the pieces together.

Chapter 1 introduces the basic concepts including the risk involved—both the risk of failure and the risks inherent in success. Chapter 2 is a short history on the subject based on a very limited amount of available information. Hopefully, there will be some readers who can provide more background on reverse engineering. I welcome additional information on the history of the subject, especially the period preceding 1950. The latter part of Chap. 2 deals with legal issues one may encounter in reverse engineering; anyone with serious questions should consult a legal professional.

Chapters 3 and 4 cover the prescreening and stage 1 evaluation and verification phases, probably the most intensive and demanding parts of reverse engineering, as well as those with the least amount of information available about how this part of the process is done. Skipping these steps are the first risks one takes in reverse engineering, and I believe the risk and cost grow exponentially if these stages are not conducted thoroughly. In any multiorganization reverse engineering program these two areas will cause the most friction among team members because these stages are so time intensive and they require someone to take personal responsibility to research an item's design past. This person would likely need to act as a resource later in the program should there be any questions about the functionality of a part, for example, as in a case when the need for specific design requirements are questioned.

Chapters 5 and 6 discuss technical data generation and design verification, which are the heart of reverse engineering. These are also the areas where there is more information than any of the others, most likely because they are the most engineering intensive and thus are the processes most engineers like best. Data collection and the economic aspects are often not as interesting to design engineers.

Chapter 7 involves project implementation, a much-neglected step in the overall process. To get a part from the design phase into the operational system is an aspect of engineering that is often overlooked. It is much easier to design the perfect widget than to make it an integral part of a system which functions perfectly day in and day out. There is very little glory to building a part perfectly, because once it does its job no one pays attention to it until it fails. Who cares that the coffee pot worked for 5 years without a problem? We only notice it the day it does not work. And so it goes with project implementation; for reverse engineering to be successful, a part has to be implemented in the system it came from to complete the circle. There will be resistance to putting the reverse engineered part back in the old system, and many of the topics covered in Chap. 7 will help with this final implementation.

Chapter 8 summarizes the complete process by highlighting a special project which required much effort in every stage. Chapter 9, Future Applications, is a jumping off point for you, the readers, to further what is presented in this text. The appendixes contain portions of very important drawing specifications. If technical data development is the heart of reverse engineering, consistent application of drawing practices and interpretation of data are necessary components of success. I cannot stress enough that we all need to be reading from the same sheet of music whether we are collecting digitized data, generating the CAD drawing, or machining the actual part. As we move toward global standards it is particularly important that we find practices that will be interpreted the same whether a part is built in Korea, India, or Macon County, Georgia.

This text is a guidebook, a reverse engineering how-to guide in specific. It is designed to be straightforward with simple examples. It is meant to be the basis for delineating an entire process, not just the easy engineering piece of technical data generation. It is meant to be shared, and hopefully it will get dirty in a shop. We welcome any information which could expand our collective knowledge of the subject.

I issue one challenge to the reader. Those with enough technical knowledge to upgrade the systems they work on and with are being given a tool in this text which can empower them to improve these systems. If anyone who reads this book can improve one piece of their system (and our work almost always binds us to a system of some sort) with the tools of this book and chooses not to, they become part of the problem instead of part of the solution. I challenge you to become part of the solution. This work offers to all who read and act on its ideas, part of the solution to staying competitive in the challenging times we face today and those we are soon to face.

Kathryn A. Ingle

Acknowledgments

There are many whom I wish to thank for their support over the years that this work has evolved.

To Bill Dugan, who as my first supervisor gave me the opportunity to develop a program that could accomplish all its aims with a minimum of red tape.

To the many people at the Oak Ridge facilities of the Department of Energy who participated in our early attempts, but especially to John W. Smith, Joe Arnold, Russ Bragg, and Les Johnson, who forever questioned every piece of equipment I brought them and every direction I gave them. This furthered the methods which serve as the rudimentary beginnings of this book.

To Tony Moore, to whom I owe all my real knowledge (or lack thereof) in electronics. For many years this man bore with me when I could not tell alternating from direct current, though I still fear electrocution when plugging in appliances.

To the late Tom Wheeler who encouraged my first thoughts on writing this text and who introduced me to Bob Hauserman, senior editor, McGraw-Hill, and coached me in my first feeble attempts to complete the forms McGraw-Hill needed. You were always an inspiration, Tom, and still are.

To Bob Hauserman who believed in the idea and worried enough for all of us. Without your support and understanding this would never have happened; your staff was terrific even when I was not.

To the inimitable Rosie Mincey, my graphics designer and friend, without whose work this book would not appear clearly.

To my mother, who taught me how to write, and my father, whose genes turned me into an engineer—blessing or curse, you decide.

To my many friends who supported me through the development stages and listened through the days this book just would not write itself. You are gold to me.

To my many pets, any typing errors were their doing, but I know they were only trying to be helpful and loving.

To John, who pushed me in ways I would never have trod otherwise and whose support helped finish this work. Many thanks and *che bella cosa*.

And with all my heart to my beloved daughter, Abigael, who listened to it all and who always wanted me to do whatever I was capable of—and I always wish the same for her. Thanks for all your love, encouragement, and sacrifice.

List of Abbreviations

BPR	business process re-engineering
CAD/CAM	computer-aided design and computer-aided manufacturing
CASE	computer-aided software engineering
CIM	computer-integrated manufacturing
COTS*	complete off-the-shelf (item)
CTDP	complete technical data package
DD*	data development
DE*	data enhancement
EMI	electromagnetic interference
FAST	function analysis system technique
FY	fiscal year
IC	integrated circuit
LCC	life-cycle cost
LCS	life-cycle savings
MTBF	mean time between failures
PDP	preliminary data package
PDS	preliminary drawing set
PTDP	preliminary technical data package

*Used in diagrams or equations only.

PV*	project verification
QER	quality evaluation report
RAM	reliability, availability, and maintainability
RE*	reverse engineer
ROI	return on investment
SIE	special inspection equipment
TDP	technical data package
TQM	total quality management
VE*	value engineer

*Used in diagrams or equations only.

Introduction

In recent times there have been events on a global scale which have changed forever the way in which companies do business. The markets companies must compete in have become international. The sources of supply for manufacturing components have become multinational. Computers and networking have added to the ease of communicating with each other and with the outside world for the exchange of information. Because of these technological, economic, and political changes in the world at large, not only have many companies altered their way of doing business but their organizational infrastructure has changed as well. The organizations we work for no longer have simple (self-contained) departments to handle single (self-limited) issues but are matrices of skilled workers. No individual worker, department, or organization exists alone or in a vacuum.

For many companies, what was once a stable production system has been interrupted. Some of the economic downturns and belt-tightening measures have forced companies to lower their level of capital investment in equipment to produce goods. Over time, the production equipment installed often differs from the original production-line design. This type of equipment diversity has more recently led to chaos in the production systems, and this chaos represents the natural progression of entropy in the universe. Those familiar with the laws of thermodynamics will recall the theory that the sum total of entropy in the universe is constant and that systems in order tend toward disorder and vice versa, that systems in disorder will tend toward order, after a certain point of chaos. So the decline in a working production system will continue unless intervention in the form of preventive or corrective measures is taken.

No "for profit" manufacturer can afford to let a production system completely decay and will intermittently intervene to maintain the system at some fairly efficient and profitable level of production. In some cases this

1

will require preventive maintenance, which may escalate to troubleshooting and corrective maintenance, then to overhaul, and then on to system modernization before the thought of replacing the system becomes an economic necessity. All-out replacement of equipment is a last resort, to be used only when there is a return on the capital investment. The bottom line for any manufacturer remains return on investment. (A resultant significant increase in productivity ultimately means there is some long-term economic gain for the manufacturer.)

With the chances of significant capital investment in new equipment being placed into further outyears, more systems need to be maintained in their present condition for longer periods of time. There are often gaps in the technical support information needed to maintain a system built from older designs using outmoded or outdated techniques or materials, whether in a small company or a large multinational corporation. These gaps may be effectively filled by *reverse engineering*, which is one of many solutions to maintaining a system for a longer period of time. While not many manufacturers have considered using reverse engineering in the past, since it was perceived as a form of patent or design infringement, many are now being forced to look anew at this technique. There are many individual reasons for this change of heart, but by far the vast majority find that when they have lost suppliers for critical spare parts, reverse engineering may be the only alternative. The loss of a supply source leaves few choices when a replacement part is not immediately identifiable. If a part has a simple substitute, then there is no real long-term problem. It is often when a manufacturer is faced with long-term issues that reverse engineering becomes a viable alternative to having no future parts to maintain a production capability.

Reverse engineering may also surface when a company lacks adequate technical data to repair its own equipment. Many systems in use today are 30 years old or, as is often the case with older pumping or power equipment, some systems are potentially 50 years old—or even older still. Developing and emerging nations are faced with similar problems of design and technical support. Even developed nations sometimes find themselves without the material needed to technically support the equipment; for example, during the Gulf War in 1990 Kuwait suffered difficulties after bombing destroyed many technical data storage facilities. Natural disasters, such as earthquakes, fires, and floods, can also cause the loss of critical, irreplaceable technical data anywhere in the world at any given time.

Today the economic viability of reverse engineering is being reconsidered. The focus on short-term, easily achievable bottom lines as opposed to long-term investments, positive returns on investment over time, and dwindling research and development efforts have together formed a continuing trend which jeopardizes the industrial base of many countries. The industrial base of any organization or nation often rests on that organization's or nation's possession of technical information, depth of manufacturing base, and gross national product and trade balance. So there exists a need to protect the technical data which are already in one's possession and to expand on the available data required for operation at a maximum level of efficiency and productivity.

Now that we have identified a very real need to develop and maintain technical data, how can it be filled? By teaching those who need to know a method or technique that can be easily followed to produce a net positive result for both themselves and their companies. Reverse engineering is one such niche technology that can be accessed to fill this need.

The aim of this book is to provide the guidance needed to steer a team of technical professionals through the relatively uncharted waters of reverse engineering. Many decisions will be left up to you, the reader. You are expected to know your equipment better than I do. You will be expected to extrapolate on the methods suggested to suit your own needs. You will need to tailor the information presented so that it fits the equipment and systems with which you are working.

This book is intended as a how-to guide to lead you through some examples of what to do and, in some cases, what not to do. It will point out some of the pitfalls which may be overlooked early in the processes and stages. It will urge you to think through many aspects of the design process that you previously had relegated to a back burner in your desire to complete a project. With any luck you will never be able to look at even the simplest pieces of equipment, such as the humble telephone, in the same way again. For every time you have a better idea you will consider all the positive design aspects before condemning an entire product. And for every time you have had a bitter experience with a real lemon you will also roar. The roaring will die down when you realize that you have the power and skill to change this ugly duckling of a design into a graceful swan. This also places you in a responsible position to enlist the assistance of, request the participation of, and empower those who have the capability to change the old into the new. With so much change in the winds of our time, if we are not moving forward, we most surely will be held back.

Before learning how to implement reverse engineering, one must understand what it is. A detailed definition can be found in Chap. 1. Generally speaking, reverse engineering is a series of four interdependent stages, each stage building on the findings of the previous stage preceded by a prescreening process. The whole process progresses linearly with time. It is charted on the flow diagrams which visually illustrate each step of each stage.

Webster's Tenth New Collegiate Dictionary defines *methodology* as a particular procedure for consistent inquiry in a particular field or discipline; thus reverse engineering is a methodology. If a *technique* is a method to accomplish a desired aim, then reverse engineering is a technique. If a *technology* is a scientific method of achieving a practical purpose, then it is also a technology. *Technique* and *methodology* can be used fairly interchangeably and still be an accurate reflection of the subject matter, but *technology* is a better description of the prescreen and four-stage processes.

It is a technology in and of itself. There is no other form of data development like it. Reverse engineering requires a synthesis of hardware and software, electrical and mechanical, metals and ceramics, and design and integration of such a variety of subjects that there is nothing else like it. *Value engineering* is a special subset of reverse engineering closer to design upgrade. The differentiation between reverse engineering and its similar counterparts of re-engineering, concurrent engineering, and software reverse engineering will be discussed in detail in Chap. 2.

So why is reverse engineering important? It is a form of technology transfer. Technology transfer is a win-win proposition for the possessor of the technological knowledge and the receiver of the knowledge gained. Sharing technological knowledge is good for the progress of the entire human race. Information hoarding no longer makes one a leader or expert. In these days of intense technological development, the cutting edge can be maintained for only a short while until a newer development ousts the leader for attention, importance, and funding. Meanwhile, the company that can make the best *use* of the newest developments can maintain the industry lead far longer in a practical sense. In my personal opinion, therefore, it is more important to share knowledge and develop better methods of utilizing the knowledge than to hoard knowledge for one particular application.

Whom is this book for? After researching all the areas I was not familiar with, I felt that I had read as much "technobabble" as I could stand. I am not out to win literary praise for this work. I am attempting to write a how-to manual which can be used by anyone with the necessary resources, from undergraduate engineering students and garage inventors, to line operators and plant managers, as well as farsighted captains of industry. It is written by a person with an engineering background for others with similar engineering or technical backgrounds. Presumably, it is in a language we all can understand.

As few acronyms and abbreviations are used in this text as possible. Minor exceptions are made for the sake of brevity to eliminate the redundant use of long phrases. Some examples are RE for *reverse engineering* and ROI for *return on investment,* used primarily in diagrams and equations for brevity, and some localized usage of chapter-specific acronyms. Abbreviations are used for industry [ASME, ASNL (American Society of Mechanical Engineers and American National Standards Institute), etc.] and government standards [DoD, MILSPEC (U.S. Department of Defense, Military Specifications), etc.] because they are cited so frequently in our everyday work. Overall, the use of acronyms has become widespread to those immersed in many types of systems. The extensive usage of acronyms can become confusing, especially to those not directly involved in work on the system. Although many people use acronyms as a verbal technical shorthand or to demonstrate a technical linguistics fluency, acronyms are often used as a way to exclude those unfamiliar with the system. Engineering and technology, per se, are not just for the privileged few with the "technically correct" education but must be accessible to all who have inquiring minds. The actual practice of design engineering is, however, best left in the hands of those with the requisite background. This statement is not meant to exclude the inventor but to enlist the practitioner of engineering in the process to ensure the safety of the public, which is the highest objective of engineering.

There are examples to guide you through calculations and real-life examples which serve to illustrate; however, there are no exercise problems at the conclusion of each chapter. This is because each real-life situation is unique and not all factors can be known or extrapolated to other situations. Reality does not match the book in this regard. The book will assume that there is more information available; ideally, all information is available.

In this regard, students can rejoice; there will be less need for homework. The flip side or tradeoff is that you will have to think harder when approaching a sample problem because only certain quantities will be known; thus, the answers to the problems will be more complicated. Because there are no answers at the end of this text, you will have to use your best judgment—kind of like real life, where there are no certainties that your choices are right. Note, however, that if you begin to get comfortable with the unknowns and assumptions of daily life, you will probably be well suited to the world of real machines with contrary personalities and many interrelationships.

Students can use this process after taking a design course or two to understand what not to do when designing their own universe. Reverse engineering can be used as a design analysis tool by students. A potential semester assignment to choose a simple gadget with few moving parts and reverse engineer it would teach the student to define an assembly/disassembly procedure, and to measure and develop engineering drawings complete with basic test procedures. A final comparison of good and bad design features as well as suggestions for value engineering the part would make a great design project.

The line operators and technicians probably possess more direct and intimate knowledge about a particular manufacturing line than many of the original design engineers. These operators and technicians know exactly what it takes to make this baby hum. Given half the chance, they would just as soon set things right once rather than repair it 100 times. The garage inventor is the type of person who cannot resist the temptation to tinker with the equipment already on hand. The inventor is also the kind of person who *always* wants to make whatever it is work better.

The plant manager would prefer some of the discussion sections of the book to the step-by-step methodology of such topics as selecting candidates, measuring surfaces, and testing prototypes. The discussions give an overview of the processes involved and the resources needed to accomplish reverse engineering projects. This knowledge is critical to making the necessary resources available to the practitioners of reverse engineering. The plant manager may initiate a reverse engineering program within a manufacturing plant and may need to be familiar with an overview of all stages and procedures.

The farsighted captains of industry will need to read both the discussions and the methodology to apply this technology. These individuals not only can bring this technology into their own industry but also have the power to bring their own industry to other parts of their country or to other countries. Their unique overview of an industry can be especially advantageous for targeting pilot efforts. All truly effective programs must have the blessings of the top of the corporate "food chain" but will ultimately be fully implemented from the lower echelons of the traditional corporate structures.

The focus of this book is on reverse engineering as it applies to manufacturing and hardware systems. This is partly because that is the origin of this author's knowledge of reverse engineering and partly because it applies to such a variety of mechanical and electrical production systems. The same principles can be applied to the service industries and probably to human services, although this should be done with great care when applied to the processing of people and not paperwork or blueprints. Extrapolation is a measure of your imagination and creative powers. To enable you to adapt the basic processes presented in this work to your own needs is the goal of this work.

In summary, remember a key concept: Reverse engineering is not a reactive tool to use on an old problem but a proactive methodology to a present challenge. This challenge is to maintain and/or improve today's manufacturing capabilities efficiently and effectively using yesterday's machines before tomorrow comes.

1
Basic Concepts in Reverse Engineering

The Problem

Reverse engineering is essentially the development of the technical data necessary for the support of an existing production item developed in retrospect as applied to hardware systems. Technical data is critical to the smooth and continuous operation of any production or manufacturing facility. Why would technical data development, via reverse engineering, be conducted after products have been produced by existing production lines? In many cases sufficient or current technical data is missing, inaccurate, or outdated. Often unavailable technical data needed to maintain and repair equipment was never furnished or purchased. This lack of adequate design information is a global problem that plagues companies of all sizes in all countries. It is neither nation-specific nor product-specific. The aim of reverse engineering is to increase productivity through improved documentation.

The following examples illustrate the need for reverse engineering.

Example. Forty years ago a production plant purchased Widget 100 as original equipment from Acme Suppliers and installed it with a lifetime service contract. The original manufacturer of this equipment was Alpha Company, who selected Acme Suppliers to distribute its wares. Alpha Company has been out of business for the last 20 years. During the first 10 years of this period the Bob Corporation, which bought out Alpha Company, repaired and maintained former Alpha Company equipment. In the turbulent 1980s the Bob Corporation was part of not one, but two, leveraged buyouts, and the present owner of the original equipment design, Capital Crooks Inc. (CC Inc.), saw this lifetime service agreement as a losing proposition.

7

You, the production plant manager, still operate the Widget 100 machinery fairly effectively; however, now you have neither the technical data to repair it yourself, the trained technicians who could jury-rig a solution, nor the capital to purchase the new Zinger 1000 that CC Inc. is trying to force you to purchase to replace that old Widget 100. Both the lack of maintenance data and the pending obsolescence of the existing equipment illustrate the potential advantage of reverse engineering as a welcome economic and manufacturing proposition in such a situation.

Example. The Wondrous Doodad Corporation (WDC) has appointed you to lead a team of multidisciplinary engineers to revitalize a doodad manufacturing plant as a joint business venture with a leading, newly privatized, Eastern Bloc doodad manufacturer. Upon initial inspection in the home office, it was quite obvious that this doodad could use some quality improvements. You have 15 years of development experience in doodad manufacture, and actually hold an early doodad patent. You lead a crack team of industrial experts in advanced doodad production. WDC has invested heavily in the "mother of doodad" manufacturing capability for 15 Eastern Bloc countries, and their top management wants you to help them revitalize not only their doodad manufacturing capabilities but also the entire doodad market in these 15 countries so that the Eastern Bloc version of doodads may become the de facto standard throughout Europe. This is a golden opportunity and all is gleaming, awaiting your magic touch until . . . you learn the depths of corruption propagated by the previous regime and the total lack of technical information to either manufacture or repair the existing equipment bought from the Chinese 25 years ago. Your technical staff cannot read Mandarin Chinese repair manuals. The one old man who kept all the technical secrets for this equipment died last spring. The ISO 9000 registrar is due to visit in 2 months. What's one to do?

Example. Here is another scenario that bears mention. In this case the original "modern" equipment downtime is excessive in the eyes of the current users or owners. The existing repair contract works, but the cost to maintain the system is skyrocketing, and you are now constantly being overcharged for seemingly all too often "faulty" equipment. You suspect that substandard parts are being substituted during the service calls but the repair person on site probably knows nothing about the quality of replacement parts. Maintenance costs are being monitored by headquarters. Profits are down for the sixth quarter, and the pricey models coming from this once touted and honored production facility are now causing heads to roll. You must cut equipment downtime to keep your product viable and profitable, or it's your turn on the unemployment line.

These realistic yet exaggerated hypothetical cases exemplify some of scenarios that can lead to the development of a structured reverse engineering methodology to develop technical data for manufacturing and/or maintenance purposes. Success in reverse engineering can be achieved in situations where data was never developed previously, such as with many obsolete components, or wherever there is the need to maintain costly, labor-intensive, or outdated equipment. Increasingly, companies find themselves with much outdated equipment in inventory (witness the electronics and computer fields) and with requirements for technical documentation from customers or regulating agencies that was never developed, submitted, or even necessary before today. "Compliance" and registration issues have also generated an increased need to document any methods or equipment not examined in detail previously.

What Is Reverse Engineering?

Reverse engineering is a four-stage process in the development of technical data to support the efficient use of capital resources and to increase productivity. The stages, all of which are conducted after a rigorous prescreening of potential candidates, consist of data evaluation, data generation, design verification, and design implementation. This process is typically applied for the improvement of production lines or manufacturing capabilities. Ideally, groupings of parts by system or subsystem produce the best pool of candidates for reverse engineering.

Accurate data development for long-term maintenance and support of a technical capability is the cornerstone of reverse engineering. This process provides a level of technical support. Since reverse engineering requires the investment of capital, reverse engineering projects are carefully prescreened to ensure a high probability of success. Those projects which do not meet prescreening criteria are typically not considered, because success in reverse engineering is generally measured by return on investment. Success in reverse engineering is also measured by overall effectiveness to both

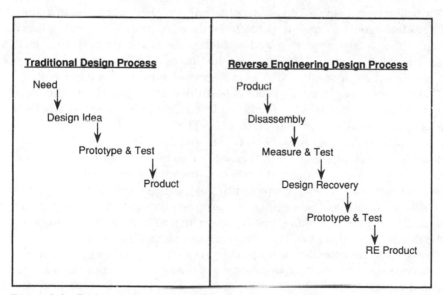

Figure 1-1. Traditional versus reverse engineering design process.

long- and short-term objectives, not merely a singular bottom line. Figure 1.1 illustrates the difference between the traditional design process and the reverse engineering process.

If *forward engineering* is the traditional process of moving from high-level concepts and abstractions to the logical, implementation-independent design needed in a physical system, then reverse engineering is the design analysis of the system *components* and their interrelationships within the higher-level discrete system. The goal of reverse engineering, then, is increase manufacturability and improve documentation by uncovering the underlying design. This design maximization process is a form of *value engineering*, which is a subset or by-product of which is component duplication with the added value of documentation and ease of future fabrication and manufacture. Reverse engineering thus is useful in that can be substituted prior to an expensive overhaul or system modernization and does not have the corresponding price tag.

How can reverse engineering really help? What are the advantages of using reverse engineering? How does it differ from the tools in my present engineering tool kit? Do I have to get rid of what I already know? Is this another management fad? What is reverse engineering?

The Reverse Engineering Process

According to *Webster's Tenth Collegiate Dictionary,* a *process* is a series of actions or gradual progressions conducted to achieve an end. In reverse engineering, critical analyses are required at each stage of the process since each stage builds on the results of the previous stages. The desired goal is always a product or component of higher efficiency or quality for a lower cost.

The reverse engineering process identifies and strengthens the weak links in any system. The identification of potential candidates involves technical research as well as critical analysis. The study of repair data history and technical drawings must be as thorough as possible to avoid poor candidate choices and wasted resources, both human and monetary. Improvements to an entire system are reached incrementally starting with the poorest performers moving upward to any potential special projects. Success for an entire reverse engineering program is achieved one project at a time through clear thinking, good judgment, and hard work, i.e., sweating the details. One major project failure can negate the effects of 10 successful efforts. New documentation support for equipment and improved system maintenance are important by-products of the reverse engineering process. These are not explicit benefits but become a subset of the technical data package development.

Technical Data Development

Accurate technical data development is the essence of reverse engineering with accurate data development as the cornerstone. This data can be in the form of engineering drawings, equipment specifications, performance characteristics, special tooling, or any other information critical to the ongoing performance of the manufacturing capability. Background information can be found in technical manuals, repair data records, performance criteria, and other vital information which can augment the engineering drawings. Candidate screening requires a thorough investigation of all available information. Since the four reverse engineering stages are interdependent, the accuracy and completeness of data at all stages is crucial, as decisions based on incomplete or inaccurate data at any one stage can be disastrous at any of the subsequent stages. In an era of resource conservation it is unconscionable to conduct partial research in any phase of data development or improvement. One does not develop data which already exist.

Definitions

Although reverse engineering ideology and theory have not been examined in detail yet, some specific process elements are introduced and defined at this point. Most of these topics will be addressed specifically in later chapters.

Reverse engineering begins with components which have been prescreened against specific criteria. *Components* are singular parts, such as a bushing or circuit card, or the smallest complete units of systems, such as valve assemblies or electronic modules. Prescreening takes place prior to the four-stage process on a "candidate" or potential reverse engineering project. After receiving a positive prescreen and stage 1 report, a candidate becomes a reverse engineering *project*. The induction of many projects becomes a reverse engineering *program*. Figure 1.2 illustrates this linear progression.

The reverse engineering project manager or team has on hand all available prescreened technical information in the form of detailed drawings or specifications and failure data and, having reviewed the economics, logistics, and technical complexity in stage 1, is now ready to determine the *project type*. The three project types are

- Product verification
- Data enhancement
- Data development

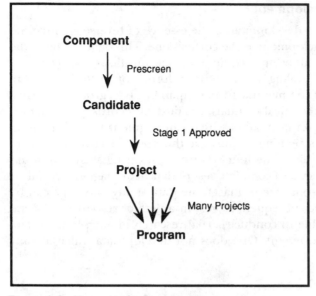

Figure 1-2. How parts develop into programs.

Product verification is the simplest form of reverse engineering; data development is the most engineering-intensive. Sample parts are procured or borrowed from the supply system. Once the samples arrive, the four-stage process begins:

- Stage 1: evaluation and verification
- Stage 2: technical data generation
- Stage 3: design verification
- Stage 4: project implementation

Project success, as discussed earlier, will typically be measured by return on investment. Figure 1.3 provides an overview of the reverse engineering process.

Figure 1.4 shows an example of a typical reverse engineered circuit-card assembly.

The original equipment manufacturer model on the upper left of Fig. 1.4 costs $625, while the reverse engineered model on the lower right costs only $450, representing a unit cost savings of 28 percent. This item cost approximately $12,000 to reverse engineer and is expected to save almost $50,000 over a 14-year period. Figure 1.5 illustrates an example of a typical mechanical reverse engineering project. The tube bundle shown in Fig. 1.5 is the reverse engineered model. While the original part cost $2922, the

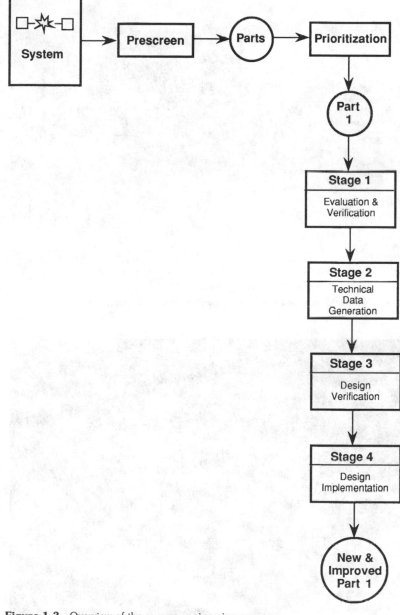

Figure 1-3. Overview of the reverse engineering process.

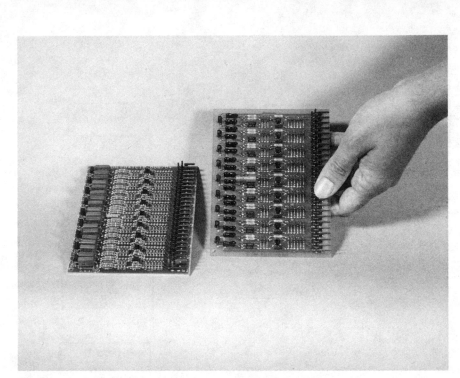

Figure 1-4. Example of reverse engineering project: circuit-card assembly.

Figure 1-5. Example of reverse engineering project: tube bundle.

reverse engineered part cost $2500, representing a 14 percent unit cost savings. The entire project, including the expense to build and hydrostatically test the prototype, cost $6556. A calculated life-cycle savings of $52,750 represented a return on investment of 7:1 after the reverse engineering cost was deducted from the life-cycle savings.

The Benefits of Reverse Engineering

With a structured methodology in place to develop technical data to support hardware equipment systems, some tangible and intangible benefits may be derived from the reverse engineering process. First is the increased ability to maintain a manufacturing capability at its peak performance rate. This is due to the improved documentation, developed in stage 2 of the reverse engineering process, for logistically unsupported equipment and systems. Once a system is maintainable, the manufacturability of the end product, which is the output of the improved system, can be enhanced and improved on through future value engineering efforts. Overall, these two factors of maintainability and improved manufacturability tend to drive the potential to realize actual life-cycle savings through the use of the improved product documentation. Any real (as opposed to estimated or projected) savings can then be used to increase the profitability of the system, and the potential profits can be pumped back into system upgrades as newer technology becomes both available and affordable. In situations where system modernization is too costly or capital-intensive, reverse engineering can function as a stopgap measure to increase system productivity until enough savings and/or profit become available to conduct full system modernization.

Reverse engineering should not be confused with *system modernization,* which involves the technological upgrade of an entire system to eliminate many portions of a current manufacturing or fabrication capability and replace these with more sophisticated equipment. System modernization encompasses many cost-prohibitive choices which capital-poor companies cannot afford. Reverse engineering can, however, buy enough time in the market cycle to pool the necessary resources to complete a system overhaul or modernization later in the future. Reverse engineering is above all targeted at modernizing singular components of a system, not the system as a whole, to maintain or increase system productivity.

Reverse Engineering as a Quality Function

Quality in manufacturing is achieved by painstaking attention to detail at every step of design, preproduction, production, and postproduction, including product enhancements. Because quality is achieved in the reverse engineering process by enlisting every employee from machine operators to the highest-ranking corporate officers, all employees can achieve quality through proficiency in their respective roles in this process.

Reverse engineering employs *total quality management* (TQM). The principles of TQM are embedded in reverse engineering practice. TQM is an approach to managing work based on the analytical evaluation of work processes, and the development of a corporate culture which can support employees empowered to impact the final quality of a product. As a requirement, every component which is reverse engineered will conform to current quality assurance procedures for both the drawings and the final technical data package, which includes procurement information that has quality assurance requirements. Quality is built into a reverse engineered component, not inspected in at the completion of the final stages. Quality assurance requirements range from ISO (International Standards Organization) 9000 for international corporations to MIL-Q-9858 for certain United States military components. Still the ultimate goal in the reverse engineering process is continuous product and service improvement.

Reverse engineering is not done in quality circles but by knowledgeable teams of competent experts from the shipping department to the sophisticated polymer coating chemical engineers and by other specialists as yet unknown. It requires systems thinking because each component is merely one piece of a larger system. Human factors cannot be overlooked. Just-in-time procedures can still be implemented. Concurrent engineering practices can be used throughout the process and embedded into the technical data package. Excellence is still a goal, as reverse engineering involves critical resource allocation in today's highly competitive marketplace.

Uses of Reverse Engineering

How will this information be of use to you, the reader? It will lead you on the step-by-step use of a process that can be tailored to the individual users' needs in production-line efficiency, customer service improvement, and other areas. It can help the manager allow system users to identify and improve system weaknesses. (Those interested in the mechanics of the reverse engineering process are referred to Chap. 3.)

Most of this book is directed toward manufacturing applications. The original applications of reverse engineering were oriented to production systems and will be continued in this vein. Hypothetical manufacturing examples will be used because it is easier to visualize the transformation of raw materials into finished products—particularly the mechanical or electrical parts of a whole system—than the service industry process of customer need to customer fulfillment. Most systems, whether human or manufacturing, can use the reverse engineering process to increase efficiency and productivity in a postproduction situation with existing systems.

Extending Reverse Engineering to Value Engineering

Value engineering is an enhancement of the basic reverse engineering process. Both processes are identical to a point, except that the goal of value engineering is to build the elusive better mousetrap. The value engineered product has the original functionality but is either smaller, faster, cheaper, easier to use, more effective in operations, or on some level a better component than the original. The new part has *added value*. This is conceivably the desired end result of every reverse engineering project; however, this text is designed to delineate the basics of a general reverse engineering process, not the nuances of the value engineering process. The degree of project variation is so high that it would be almost impossible to write a definitive text for all types and cases of value engineering, and it is often difficult to determine exactly when a reverse engineering process becomes a value engineering process.

Value engineering is an added value, a significant improvement, a new look. Value engineering can be accomplished by applying newer technologies, materials, or techniques to improve the quality of the part or to cut the costs of production. Simple design substitutions during reverse engineering do not always constitute value engineering. An example of simple design substitution which does not qualify for value engineering would be to replace an older type of resistor with a newer one that is listed in electronics supply catalogs as a valid replacement part for the older resistor. The case of substituting commonly available punched sheet metal on a strainer for the older hand-drilled sheet may, on the other hand, constitute a significant cost savings and therefore may qualify as a value engineering solution.

Customer feedback has been in vogue for only a decade or so, and many older designs have remained unobtrusively in the background of a system, without being questioned or tested for efficiency, while newer technologies have surfaced. In value engineering, it is important that the design functionality and capabilities be questioned. This questioning, particularly if customer or user feedback is available, creates lateral and vertical room for design improvement, and true value engineering can begin.

It may be known from the start that the original component must be improved on because duplication of the original design would mean duplicating the unwelcome design flaws as well. Value engineering may not become an appropriate or apparent goal until stage 1, 2, or 3 or even in some cases until stage 4. It is necessary to always be alert to situations where there is no room for improvement, where the original design will not tolerate even minor changes, as well as areas of potential improvement.

Value engineering, as another methodology, often focuses on improvements through cost reduction. Other areas such as customer-perceived quality and/or performance requirements or increased functionality also add to the value of the end product. Improvements can also be obtained through new manufacturing processes and procedures. Value engineering can be applied not only to hardware systems but also to software, processes, procedures, organizations, and computer systems. Value engineering is often applied to substandard products to improve the quality or lower the cost as added value.

Reverse engineering and value engineering are very similar in nature and in application to systems; their main distinction lies in the end product. The goal of reverse engineering is to create a duplicate component, as an exact one-for-one match, while the value engineering process is taken a step further. In value engineering the goal is to improve the end product while duplicating the functionality. The value engineered component may cost less, weigh less, be smaller in size, and still perform the same function within the system. It is also possible to value engineer a whole system, but only specific components would be useful to reverse engineer. Reverse engineering a whole system would be uneconomical. In theory, applying value engineering to a whole system would be most used prior to, and during, the system design stage, not during postproduction. If an improvement technique were to be applied to an existing postproduction system as a whole, it would seem expedient to use system modernization, not value or reverse engineering. (*Note:* This is confusing only because of the similarities and subtleties. We modernize systems; we reverse and value engineer components.)

Figure 1.6 shows an example of a value engineering project. The ultraviolet test unit as supplied by the original manufacturer was used to test sensors in a gas turbine. This original unit was explosion proof and had a magnetic trigger assembly. It had a recharging unit and weighed about 10 lb. After discussions with the users it was determined that this piece of equipment was used only while the gas turbine was shut down to determine whether sensors were operational. The recharging unit was heavy, costly, and required constant maintenance as it had to be returned frequently to the manufacturer for overhaul. The unit was replaced with a lighter-weight handheld flashlight assembly that required AA batteries and less frequent maintenance. The overall unit weight decreased also. In the replacement unit, the original lightbulb was changed from a fluorescent bulb to a readily available ultraviolet bulb and a 1-in section of the protective cover was cut out, permitting the ultraviolet light source to emit lightwaves strong enough to trigger the sensors for testing. After system testing, the user felt

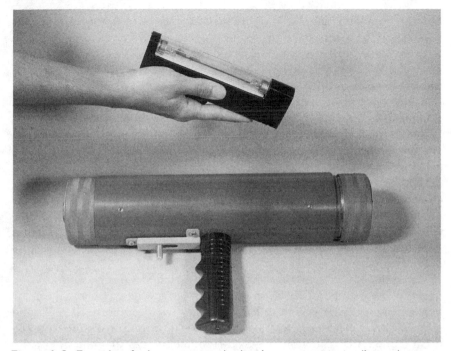

Figure 1-6. Examples of value engineering (*top*) and reverse engineering (*bottom*) projects: ultraviolet test unit.

this value engineered model worked better for testing these sensors than the original unit, which had only a pinpoint window of coherent light with which to hit the target sensor at exactly the right spot to accurately test the sensors, while the handheld model had to be pointed only toward the target area.

The original unit cost was $558, not including the charger, while the value engineered unit cost $30, with the ultraviolet lightbulb and all design changes (including batteries). This produced a 95 percent unit-cost savings. The project was estimated at $5000 and actually cost $6880. Still a life-cycle savings of $765,600 was achieved and the return on investment was 110.3:1!

Value Analysis

Value engineering is often equated with value analysis. Although the methods vary, the objectives of value engineering and value analysis are similar: to find the optimal design approach. With value engineering there are usually more severe restraints on the variations because the part has to fit back into the original design, whereas in value analysis the optimum design does not have to be system-specific. The improvement of a screw thread design does not have to be application-specific in value analysis, whereas in value engineering the improved thread design must fit the higher assembly of the specific mechanical component.

Value engineering evolved from value analysis as an implementation of value management. Some consider this to be the true predecessor to TQM and other management optimization techniques. Value analysis is a systematic method for evaluating several proposed design options. Value analysis may suggest new and infinitely better designs. *Value* in manufacturing is described as the ratio of performance or function to cost. With multiple choices for one application there is usually a breakeven point when all factors are equalized and one design demonstrates a significant advantage beyond this point. Again, value analysis need not be application-specific and therefore is less restrictive in terms of the design options which can be studied.

The roots of modern value analysis begin with the design optimization trends of the 1950s and 1960s. In the 1940s, Larry Miles created a dramatic new problem-solving system called *value analysis* to optimize the cost and efficiency of a product or process. This was improved on by the development of the FAST (function analysis system technique) diagram by Charles Bytheway in 1963 which graphically show the relationships and interrelation of identified functions within a system. Efficient and effective design production techniques and optimization schemes have been in existence for quite some time.

Value Engineering Practices in the U.S. Government

The U.S. government often uses value engineering to describe an organized effort to analyze the function of systems, equipment, facilities, and supplies to achieve the essential functions at the lowest life-cycle cost consistent with required performance, reliability, quality, and safety, although this concept applies equally to commercial organizations.

Value engineering in a government contract is used to upgrade equipment and systems within certain limits. The Federal Acquisition Regulation (FAR) Part 48 describes the scope of a value engineering change proposal as a way to reduce the cost of equipment/systems supplied to the government, along with the necessary policies and procedures, and contract clauses. Solicitation provisions and contract clauses are specified in FAR Part 52.248-1 and provide the incentive for government contractors to provide value engineering services. FAR 52.2448-1 provides a formula to calculate the instant, concurrent, and future contract savings; the contractor receives a percentage of the net life-cycle cost savings. In this manner design improvement is encouraged in the supply of goods and services to fixed-price contracts which benefit the contractor, the government, and the taxpayers, making it a net benefit for all.

The Risk of Failure

Reverse engineering, like any engineering process, is not without risk. It would be a breach of good practice to not recognize up front that not all candidates will become projects and not all projects will realize success. Since fiscal resources must be allocated for this process to begin, continue, and succeed, the risk of failure should be addressed. It is perhaps apparent by now that the entire process is fraught with danger and risk. As some cultures are aware, every incidence of danger has corresponding characteristics of opportunity. Tremendous success can be snatched from the jaws of apparent failure.

If 1 in 10 potential system components becomes a prescreened candidate entering the four-stage reverse engineering process, 90 percent of all system components remain unchanged after prescreening. Of the prescreened candidates that become reverse engineering projects, over the long term these may only realize a 65 to 75 percent overall success rate. Perhaps still only 90 percent of these successes will truly become fully integrated into the original manufacturing process. Prescreened candidates should not be entered into the four stages without a 25:1 projected return on investment, and special projects with high risks can succeed with returns of over 200:1. For every one dollar of U.S. currency (although the peso, the yen, or the mark could be substituted) invested, very real return can be realized.

Using an example system of 100 parts, this translates to only 10 parts passing through the prescreening. Of the 10, only 7 will be inducted into the four-stage reverse engineering process. The three parts not inducted will either have no real return on investment or a low priority. One, possibly two, will have to be eliminated from the four-stage process for a variety of reasons explained more fully in Chaps. 3 and 4. Only six parts will actually be reverse engineered. An average reverse engineered project will cost $10,000 to $25,000. If the cost to reverse engineer is assumed to average $20,000, the six parts will cost $120,000 to reverse engineer. With an average return on investment of 25:1 (taken from a study of over 120 parts reverse engineered by the U.S. Navy), which for this example will be lowered to a 20:1 return on investment, this will equate to a system savings of $2,400,000 for the $120,000 invested.

This proves the economic viability of reverse engineering when done in accordance with the guidelines of this text. These possibilities can lead one back to the question "What is the catch?" The catch is the risk. There is risk if the data collection is incomplete. There is a catch if the data is not evaluated with intelligence and technical expertise. There is risk associated with poor execution of the four-stage process. There is risk for every deviation from standard engineering practice, technique, and factors of safety. There is risk that no return will be achieved if the project becomes "un-reverse engineerable" in the middle of stage 3. There is risk that the part will never be brought to the new reverse engineered design. There are unforeseen liabilities and legalities with present legal practices which could change tomorrow given today's dynamic economic and legal climate (particularly in economically or politically unstable countries). In many cases there will be no more risk than continuing to use outmoded technical components or not meeting production quotas. In short, there is risk in every step of the four-stage process. Lowering the risk of failure means increasing the risk of success.

The Risk of Success

Often it is wise to invest moderately in a budding reverse engineering program by choosing only the candidates that have the highest possibility of succeeding for the smallest investment. A minimum goal of a 25 percent reduction in the unit cost of the item due to reverse engineering is anticipated. This is evidenced in the projected return on investment, which is carefully monitored from prescreen through stage 4. The return on investment is the ratio of the net savings to the cost to reverse engineer. The net savings are achieved primarily by decreasing the unit cost. If reverse engineering does not decrease the unit cost, then there is nothing to be gained. If, on the other hand, there is something to be gained, then this fact must be established. Since most good ideas need to be sold to those who would fund such ventures, it is helpful to have solid economic data which support the conclusion that reverse engineering is a valid approach to a systemic problem.

Since return on investment may take some time to realize in even the shortest-term projects, it is wise to allocate enough financial resources to continue a reverse engineering program before expecting returns to be available for reinvestment. Reverse engineering projects typically require a time investment of 1 to 3 years before returns are realized. It takes fiscal year (FY) X dollars to begin reverse engineering. The project may require 2 FYs to be reviewed, designed, tested, and implemented. This means the time investment is FY $X + 2$, or perhaps $+3$, for the return on investment. In a period of 2 to 5 years it is conceivable that a good reverse engineering program can become self-sufficient. Commitment from top management is necessary because reverse engineering is not about short-term improvement, although short-term returns can be used to continue a long-term reverse engineering program.

It may be worth mentioning now that there is an elite class of reverse engineering project called the "special project" which is a high-dollar, long-term project of a series of related items. By grouping these items together, one can lower much of the costs of data evaluation and development, along with some of the risk by developing interrelated parts at the same time. These can achieve returns on investment of 100:1, 200:1, or higher. These substantial projects cannot form the meat of any reverse engineering program but can substantially boost the return of the entire effort.

The world we live in is a complicated mix of technology, production, marketing, and economics. Companies and products no longer can exist in a vacuum. This has led to global competition for markets and resources. *Quality, efficiency, producibility,* and *foreign competition* are terms that evoke a strong reaction in every manufacturing interest in the world. There are multitudes of solutions to the myriad of operational problems for every producer of goods. From corporate philosophies which create the culture of the organization to the humblest employee, tools are used to produce portions of the finished product. Reverse engineering is a tool. It is not the sole domain of the engineer. It can be used to improve any product line, service, or capability within an organization. It is a good business practice. It can be an act of technology transfer, national goodwill, and a simple investment in the future.

2
History of
Reverse Engineering

The Origins of Reverse Engineering

It is difficult to pinpoint a specific moment in time when reverse engineering came into being. It seems more an evolution of ideas. One of the oldest ideas in engineering is building a better mousetrap, which probably followed close on the heels of design in its original state. For every design, there is a critic, and for every ambitious critic there is, presumably, a better design. Better designs are due partly to expansion of our collective technological knowledge and the continuous pursuit of a longer-lasting product.

In all the time I have personally spent doing reverse engineering projects I have never learned the origins of the term. I suspect that it was around a long time until it gained credibility as a method to redevelop technical data. Perhaps it has been around since the dawn of the industrial revolution, and hopefully it will be around long after the computer revolution and the ever-increasing pace of technological advances have become a staple for those who pursue any type of improved design.

As long as people have wanted to understand what makes things work, there have been those curious enough to tear apart that which is a mystery to them. To fully understand a design it is important to disassemble the original item and then try to put the puzzle back together. During the disassembly stage one hopes to discover the hidden secrets by finding the mechanism, or mechanisms, that make it work. Whatever it is that makes it valuable stirs the human imagination which desires to possess its secrets by seeking and finding the keys to unlock its magic.

Preindustrial scientists developed a method to consistently investigate the nature of things. This is now known as the *scientific method*. Without this scientific method many conclusions to individual investigations were inaccurate descriptions of the truth. One such example is the erroneous conclusion that the earth was the center of the universe, or that the world revolves around us. As most 2-year-olds learn, their world enlarges and they soon discover that they are not the center of the household. It is a shocking yet necessary developmental stage in the growth of any human being. After the discovery that the self is not the center, then self-mastery can begin. So it was with the earliest scientists, who needed to master certain secrets or unknowns before making things work better. Some time in fairly recent history reverse engineering became a form of design mastery. Before we can improve the designs of the present, we must discover their secrets.

It seems that industry has been conducting some form of reverse engineering for a long time. If necessity is truly the mother of invention, then initially this crude form of reverse engineering took the form of necessity as people ventured far from the shop of the tradesperson who produced some piece of equipment, and had their equipment fail. If the individual whose equipment failed could not understand the design, it could not be fixed. If it could not be fixed, it had to be either handed off to someone who could fix it or brought back to a similar craftsperson who could handle the repair. In this manner, the possessor of the design knowledge became an important figure. Eventually some of these people had insights into a better way to make this equipment. This better way became the standard until someone else came along to improve it further. And so designs improved in iterations until the art of practical design became a craft in and of itself.

Those engaged in design continued to improve their skills by backfitting older equipment, and after much trial and error the jack of all trades became an important type of support person to have when adventuring into any type of wilderness. Eventually skills crossed over artificial boundaries (of the kind built into apprenticeship programs) and one design idea could be implemented in other applications. With the outgrowth of as many technologies as we have today, we now have to study broader subjects to make better designs.

What became necessary in the wilderness, or any place where the original designer was unavailable, later evolved into forms of copying designs and in some cases design theft or patent infringement. With design theft came the need to protect original designs which is today the patent. A patent is a registered, documented, and approved original idea or design. A *patent* in general refers to the granting of certain rights by the government of a particular country. The granting of specified rights is given for the patent of an invention. These rights generally consist in the "exclusive" right to make or manufacture a patent item for a limited period of time. These rights can be made for the positive manufacture, or the exclusion of others from making, the patented item. The person granted the right of patent has a monopoly over the specific subject matter of the grant for a specified period of time.

Patents for inventions were first granted in some Italian states in the fifteenth century, and the concept spread through the European states throughout the next two centuries. A set of formal and comprehensive patent statutes appeared in the United States in 1790 and in France in 1791. The French statute is important because it explicitly declared that the inventor had a natural and exclusive right to the invention. The objective of the exclusive right was to reward the inventor and thereby stimulate inventive activities, encourage the disclosure of the subject matter for public knowledge, and encourage the production of the item. The inventor would then be free to manufacture without competition for a certain specified period of time, thus encouraging others to invest in the enterprise of manufacturing the item. This exclusionary right also served to limit competition.

The documentation of a design is critical to the granting of exclusive rights. If a design is documented, it can be protected. This documentation has become a form of design control, and wherever there are controls, there are those who wish to make a profit by controlling the design, to become the sole manufacturer. In essence this is a positive thing. With the advent of so many new technologies, this has become a barrier in many cases. When I can purchase a product from only one manufacturer, I am held captive. As long as the price is reasonable and the supply accessible, this is a fair situation. When the price is artificially inflated to produce higher profit than the value of the product plus a percentage of its value, the manufacturer-customer relationship is abused. From the manufacturer's standpoint it becomes a powerplay, and the customer is at the mercy of the profiteer. When the source of supply goes out of business, or otherwise no longer manufactures the product, those who have come to depend on this product are left stranded.

The pace of technological change has increased so rapidly in recent years that the manufacturing base of the world is in a state of flux. This flux has put many production facilities out of business, leaving many without alternate sources of supply. If the technical data necessary to bridge the gap left by a needed product from a defunct producer is not filled by the transfer of the data needed to continue manufacturing the original design or replaced quickly by another ingenious design, then a method to regenerate the design must be found. Reverse engineering is such a method.

Government Efforts in Reverse Engineering

The U.S. government became involved in reverse engineering in the mid-1980s when the public became aware of cases involving the procurement of a $400 hammer, a $600 toilet seat, and a $1100 coffee pot. At the same time many pieces of equipment were without a source of supply due to obsolescence. These two problems were highlighted during a review of the federal procurement practices for spare parts. The overpricing of spare parts was to be addressed by completing the future purchases of these and other parts through competition. This was mandated by the Defense Acquisition Regulation (DAR) Supplement 6, dated 1 June 1983, which provided uniform policies and procedures for the reprocurement of spare parts. DAR Supplement 6 was further reinforced by the Competition in Contracting Act (CICA) of 1984. The Competition in Contracting Act of 1984 stressed the importance of competition in the procurement of all goods and services for government use. DAR Supplement 6 delineated the guidelines for the services to reduce the cost of replenishment parts by evaluating the feasibility of acquiring parts by competitive procedures or by removing the constraints or barriers to competition. This represented a change from sole-source procurement of spare parts to competitive acquisition. One way to compete the purchase of spare parts was to share the design information and technical data needed to manufacture the part in a procurement data package with multiple manufacturers of like capability. The typical procurement data package contains six parts:

1. Scope
2. Referenced documents
3. Requirements
4. Quality assurance provisions
5. Packaging
6. Notes

The current guidance to the defense community is governed by Defense Federal Acquisition Regulation Sec. 217.720-2, whereby a variety of acquisition methods are used to achieve competition in the procurement of spare parts. In referring to reverse engineering, only one notation is found.

> As a last alternative, a design specification may be developed by the Government through inspection and analysis of the product (i.e., reverse engineering) and used for competitive acquisition. Reverse engineering shall not be used unless significant cost savings can be reasonably demonstrated and the action is authorized by the Head of the Contracting Activity.

From this need to competitively acquire spare parts and the directives listed above, the government's role in reverse engineering was developed. The guidance for the process of gathering, reviewing, and obtaining technical data was clearly specified in the 6-step process in DAR Supplement 6, but the guidance for reverse engineering was lacking. After 1985, government organizations and contractors were conducting reverse engineering, but there was no clearly defined process to handle a large set of candidates.

From this information vacuum emerged a U.S. Navy activity which enlisted help from a Department of Energy (DoE) national laboratory for technical assistance. The U.S. Navy and DoE together launched a successful program which processed over 150 candidates that had an average return on investment of over 25:1, with some special projects that achieved a greater than 300:1 return. Much of this book is based on the experience gained in the time this work was conducted. Because of a severe lack of reference material on the process, there was a great deal of room for creativity to build a process that did work for all candidates reviewed. Technological improvements and unending design questions have driven many of the improvements of that process to the development of the methodology described in this book. There is still room for improvement, and anyone who can further the art of reverse engineering is encouraged to do so. There are, however, many methodologies which are easily confused with reverse engineering, and there are also some subtle differences which should be clarified.

What Reverse Engineering Is Not

There are many things that reverse engineering is not. It is not all things to all people. It is not the manufacturing equivalent to the cure for cancer. It is not even an overlying process. It is a singular methodology useful to solving system-specific needs that cannot be fulfilled by ordinary means. If simple component substitutions with alternative commercial off-the-shelf equipment can be utilized, it would be foolish to pursue a reverse engineering solution. Rhetorically speaking, one should ask "Why use Waterford crystal when Flintstones jelly glasses will do?" Reverse engineering does not have global applications even when applied to an entire system. Only certain qualified, prescreened items will be targeted for reverse engineering. If the reverse engineering methodology is applied properly, the entire system within which the components operate will benefit.

There are many fashionable and similar new methodologies such as concurrent engineering, re-engineering, or software reverse engineering in use today. To differentiate these other similar processes from hardware reverse engineering, they must first be defined. Value engineering and value analysis are described in the Chap. 1 section on extending reverse engineering to value engineering and will not be reiterated here.

Reverse Engineering Versus Concurrent Engineering

Concurrent engineering includes the successful use of cross-functional teams, particularly by building a working relationship between design and manufacturing which can significantly shorten development cycles to produce higher-quality products with lower manufacturing costs. The emphasis in concurrent engineering lies in shortening the time it takes to get from the design to the production cycle.

Concurrent engineering can be conducted on a reverse engineering project by focusing on the manufacturability of the component. This would essentially produce a component of higher value with lowered manufacturing costs and would constitute a form of value engineering. The ease of manufacturability is not the direct goal of reverse engineering, but it is a goal in concurrent engineering. Manufacturability could, however, easily be embedded in the reverse engineering process.

To clarify the value and importance of concurrent engineering, a brief history of concurrent engineering is included. Beginning in the late 1940s, appropriate analysis methods were needed to improve production techniques. Today the focus is toward computerized methods for manufacturing control, especially automation, robotic interfaces, and computer-integrated manufacturing (CIM). Concurrent engineering applies a systematic approach to manufacturing along with the future role of people in manufacturing processes. The goal is to develop useful, reliable, and economical products using a centralized approach reliant on a broad range of expertise in lieu of the common distributed method wherein an isolated group of design engineers develops the product design, which is then handed off to another isolated group of manufacturing engineers left to implement manufacturing processes necessary to produce this part. Often the two groups of engineers are separated by time, function, and geography. Concurrent engineering sounds reasonable and simple until a large complex system requiring engineering expertise across a broad range of specialties is needed to produce a complicated design—say, for example, a satellite launch system.

The classical definition of concurrent engineering is "a systematic approach to the integrated, concurrent design of products and their related processes, including manufacture and support. This approach is intended to cause the developers, from the outset, to consider all elements of the product life cycle from conception through disposal, including quality, cost, schedule, and user requirements."

The latest definition places value on the team concept empowered by top management and follows the format of the poetry of e e cummings.

> Concurrent engineering is a **systematic approach**
> to integrated product development that emphasizes

      response to customer expectations
and embodies
      team values of cooperation, trust and sharing
in such a manner that
      decision making proceeds
with large intervals of **parallel working** by all life-cycle perspectives early
in the process,
      synchronized by comparatively **brief exchanges**
to produce **consensus.**[1]

Many engineers who do reverse engineering do not need to specialize in concurrent engineering, but it would be advantageous, even visionary, if components considered for reverse engineering were always viewed within the system in which they operate. Using the principles of concurrent engineering, wherever applicable, would be using sound engineering judgment, particularly if more than one component of a system were being evaluated for its potential as a reverse engineering candidate. If an entire system or production line were being evaluated for potential reverse engineering candidates, it would make even more sense to take the integrated approach of concurrent engineering to improve not only the *production* capabilities but also the *producibility* characteristics of the candidate. More information about the current state of the art in concurrent engineering can be found through the Concurrent Engineering Research Center (CERC) at the West Virginia University in Morgantown for readers in the United States or the Concurrent Engineering CIRP in France.

Reverse Engineering Versus Re-engineering

Re-engineering was once basically associated with software and computer code. Currently it applies to the global view of process redesign and is being used everywhere from corporate structures and cultures to software. When applied to information engineering software, it is the process of restructuring existing code, with no functional changes to the system. Generally a software system is converted into another (better) software system of similar functionality by re-engineering. Re-engineering is typically considered to be a subset or by-product of software reverse engineering. In information engineering, reverse engineering is applied to enhance the maintainability of a system without the need to redevelop functionality. A task involving the restructuring of a system so that changes are transparent to the end user, particularly one which involves no design analysis and optimization (quite unlike value engineering), would be an example of software re-engineering.

A new strain of re-engineering called *business process re-engineering* (BPR), although not simple in nature, involves simply the redesign of business processes to achieve improvements by the removal or considerable upgrading of processes, especially prior to automation. It involves taking a long, hard look at how one does what one does to see if there are redundant or unnecessary (non-value-added) steps, as well as steps which could be combined. By interacting with associated departments to review a business process, one can achieve substantial gains in productivity and efficiency by taking a systems view of the process. BPR is an admirable application of re-engineering to processes in lieu of the classic software application. It is hoped that the use of reverse engineering can have additional applications by extrapolation to newer uses as re-engineering recently experienced.

Difference between Hardware and Software Reverse Engineering

Hardware reverse engineering differs from computer software reverse engineering in many ways. The primary target in software reverse engineering is the computer code, or software, of a computer program or system. *Software reverse engineering* concerns the extraction (recovery) of higher-level design or specification information from the computer software. The construction of software for computers is in many ways similar to the construction of hardware, but the design processing hardware has to account for the materials used, the limitations of these same materials, the conditions under which the component must operate, the mean time between failure, and the cost or size restrictions imposed by external factors. In software construction, where the material is the language used, the types of applications for which the language is most useful and the operating conditions impose limitations on the code, the reliability requirements may be as stringent as the mean time between failure, and the time to complete an operation may be as restrictive as any cost constraint.

Among the most common reasons for software reverse engineering are the needs to

1. Understand design more easily.
2. Transform old systems into a current computer-aided software engineering (CASE) environment.
3. Allow for the management and reuse of existing process and data models.
4. Change technology and vendor requirements.
5. Understand the current design sufficiently to allow its redesign.

The obvious difference between hardware and software reverse engineering is simply the objects of the process, not the objectives. While software reverse engineering concentrates on computer code, hardware reverse engineering is concerned with the parts of a manufacturing or production-level system. Many good texts detail the steps involved in software reverse engineering, but this text will not overlap those topics.

After some investigation it appears that software reverse engineering is a more mature art form than hardware reverse engineering. This can be accounted for by the higher rate of change in computer systems, languages, and applications. Besides the faster growth rate, software systems have had much less tradition to contend with than have hardware systems. This faster growth rate has, in turn, spurred the need to redesign existing software systems.

It is important to note that the line of distinction between hardware and software has begun to blur. In today's environment many hardware systems are controlled by software, and much software is embedded in hardware. The two commodities appear to be merging into integrated systems that will become inseparable in the future. This will hopefully increase the use of reverse engineering as a tool of improvement without renovating the systems entirely.

Legal Issues

Certain legal issues are of concern in reverse engineering projects. Few of these issues have been fully challenged in either domestic or international courts of law, and there are some areas in which the laws are changing. This section is not intended as a full discourse on these issues, but only to mention that it could be costly to ignore the legal issues that might arise during a reverse engineering project.

Patent Infringement and Theft

Patent infringement and design theft are the primary legal concerns in reverse engineering. Reverse engineering, in and of itself, is not patent theft. If a system component proposed for reverse engineering is patented in any country, then all reverse engineering efforts for that component must be discontinued. If a portion of a component is patented, its use within the assembly should continue. Continued use of a patented part is virtually inviolable. Any illegal infringement on a patent is a criminal offense and is in no way advocated as responsible reverse engineering. A discussion of the restrictions on a patent can be obtained by contacting the patenting body in the county where the patent is registered. The rights and restrictions vary from country to country, so do not assume that familiarity with general guidelines is sufficient in all cases.

Reverse engineering is a legal practice in most instances and is not a form of design infringement or theft. Any unpatented design can be reverse engineered, although guidelines for trademarks, tradenames, and copyrights are less clear. In many cases unless it is economical to complete the reverse engineering process and then remanufacture a part which is significantly less costly, reverse engineering is impractical and unjustifiable. To do a full-scale reverse engineering project is often an expensive proposition. To do reverse engineering in order to steal another's design is neither ethical nor professional. (*Note:* Gross overcharging by someone who has managed to corner the market on the supply of a particular good or service is equally unethical.) Design infringement is not the intent of reverse engineering. Technical design documentation for maintenance and supply support is the desired end. Reverse engineering is most useful when design information is critical to the continuing and efficient operation of equipment or production facilities.

Since much of international law regarding technical design and ownership issues is changing, it is best to consult a corporate attorney specializing in patent and/or international law if any questions arise during the design evaluation phase. This is not a small matter which can be postponed until too many hours of labor have been invested.

Proprietary and Restricted Data

Another potential legal issue can be the use of proprietary or restricted technical information. Usually *restricted data* means that you, the reverse engineer, cannot access or in any way use this information in the reverse engineering or value engineering process. Again, when these types of information arise during the reverse engineering process, they should be addressed on a case-by-case basis and may require the need for legal counsel.

Substandard Parts

There is another danger in reverse engineering: a substandard part such as a bolt or fastener made below required quality standards. To reverse engineer, not value engineer, and promulgate the replication of a substandard part could be disastrous in systems which cannot withstand the large tolerances and lack of safety factors common in substandard parts.

Restrictive Contractual Clauses

There may also be restrictive contractual clauses in the procurement of special items which prohibit any practices such as reverse engineering. Warranties can be voided by the use of reverse engineered parts in some cases.

Foreign licensing agreements may also be involved. There are sometimes potential health and safety hazards. There are legal statutes involved or other factors which would be unwise to ignore. There are patent and licensing rights to observe, also. Reverse engineering is not advocated in cases where another's rights or laws are being violated. Your rights stop where another's begin. All unethical practices are to be avoided along with explicitly illegal actions.

References

1. K. Joseph Cleetus, CERC-TR-RN-92-003, Concurrent Engineering Research Center, West Virginia University, Morgantown, West Virgina.

3

Prescreening and Preparation for the Four-Stage Process

Reverse Engineering Teams

Reverse engineering is a multidisciplinary collaboration which requires the talents of many types of engineers, technicians, draftspeople, and shop personnel. There is the need for estimators and production-manufacturing workers as well as a wide variety of experts in metallurgy, circuit design, vibration analysis, ceramics, or any other specialty the inducted candidates require throughout the process.

This multitude of personnel will be needed to form the entire reverse engineering team; however, only a fraction of these people will form the core team. The core team will not change from project to project to maintain a consistency in the application of the reverse engineering principles governing the way projects are handled and will learn with each project, and this accumulated knowledge will not be lost from project to project. Multitalented individuals offer a particular advantage in their ability to synthesize the variety of elements needed for the range of reverse engineering projects that a mature program will review over a period of time. On average, components of an adequate system should be 80 percent reliable. Of the remaining 20 percent of components a maximum of half will probably be considered for reverse engineering. Of the 10 percent ineligible for consideration, most will be rejected for economic reasons. A system with less than 80 percent reliable components is actually an ideal place to find reverse engineering candidates. The core team will need to be dedicated to the tasks that are offered for reverse engineering, while other personnel are matrixed to or on loan from their specialty areas.

A fairly complex project will require the skills of upwards of 20 people from across a facility. This team may consist of one or two people involved in prescreening with two others assisting in data collection, a lead engineer, one or two design engineers to define the project and set the direction for the project, a financial expert and an assistant to handle database requirements such as tracking, a project estimator, a dimensional inspector, a metallurgist or material identification specialist, a welding engineer, a draftsperson, a drawing checker, a person who can approve the expenditure of funds to do the prescreening and reverse engineering, shop personnel to build and test prototypes, and inspection personnel. With so many skills being required it is easy to understand the financial risk of expending funds if no return on this investment can be identified. By reviewing this list it is also easy to understand why the multitalented individual is such an asset. Note that each person is required to assist for only a percentage of the project. For example, the lead engineer will be dedicated to the project but only to oversee the work, not actually do the inspection portion. The financial expert will need to review the charges to this account only on a weekly basis to guarantee the accuracy of the total project cost. The drawings will need to be checked only a few times before admission to the next stage where others take over the project responsibilities.

Members of the prescreen team may, from time to time, cross over to serve on reverse engineering projects. The prescreen members should not also form the core team. This practice is recommended because there is an obligation to review *all* available technical data during the prescreen. Often there is technical information with restrictions or limited rights for use. While this information can be reviewed in the prescreen, it cannot be forwarded to stage 1 and the core reverse engineering team. If the same person were to serve on both prescreen and core teams, with the opportunity to access restricted data and carry that information over to stage 1 and beyond, a conflict of interest might arise. Rather than limit the duties and responsibilities of a person with access to the restricted or limited data, it is easier to avoid the possibility of such conflict by separating the team members' responsibilities.

Personnel will vary for each individual project and each stage during the process. Core reverse engineering team members will not change from stage to stage or project to project; however, "virtual" (temporary; for duration of the project only) reverse engineering teams will be needed to work on projects which require specialty skills.

Engineering and Technical Team Requirements

The formation of a small, yet eminently qualified, core reverse engineering team is best accomplished with a pool of experienced talent drawn from engineers, draftspeople, metallurgists, and material identification specialists as well as lead shop personnel. A lead program engineer typically oversees the prescreening and also the four stages of the reverse engineering process. The best-qualified individual to lead the core team is most often a generalist with some background in mechanical, electrical, industrial, process, and/or manufacturing engineering. This person will be the interface between management, approval agents, and the core team, will access shop personnel and work with users when additional input is needed, and thus should also have excellent communication skills. Specialty engineers are required on a case-by-case basis depending on the types of components selected for reverse engineering. The lead engineer is required to integrate the information these various contributors produce.

Experienced draftspeople are indispensable. Assumptions about tolerances, interference fits, performance requirements, or the finer points of design must be questioned as early into the process as possible. An experienced draftsperson can raise these serious questions early in the design review and technical data generation phases. The metallurgists and other specialists with expertise in surface coatings, vibration analysis, welding techniques, and similar areas are invaluable to the projects that require these skills. Shop personnel are never to be overlooked in this process. Unless the lead engineer can operate a coordinate measuring machine or numerically controlled tooling single-handedly, it is best to enlist shop support early on. The lead engineer should also be the liaison to the shops to provide a direct interface with these critical support personnel. No reverse engineering project can be completed without some shop support. Shop support is needed in the disassembly, measurement, test, and inspection portions of stage 1 and is critical to the prototype testing in stage 3; therefore, good rapport is essential and early team involvement can avoid later confusion about the direction and purpose of any given project.

Just as virtual corporations are formed to meet specific customer needs which are later dissolved when the need has been fulfilled, virtual project teams should be formed to address one-of-a-kind reverse engineering projects. Each virtual project team is a matrix of talent which focuses on a singular project or related group of projects for a short time and is later disbanded.

Communication with System Users

Many design questions arise during the course of a reverse engineering project. The available data is often insufficient, which leaves room for many assumptions, and additional data must be obtained to make as educated a decision as possible. Developing a good rapport with the system users is crucial to the resolution of many thorny design issues and aids in making many decisions wisely. Reverse engineering can be done in an information vacuum; however, a better final product can be obtained with less risk if there is communication with the system or equipment user. Frequently manufacturability issues can be answered only by the end user. An example might be a large filter with an original design which has hand-drilled holes in a $\frac{1}{4}$-in steel plate which is then formed to the body of the filter. Hand drilling is a labor intensive and therefore expensive process. Can predrilled circular tubing replace the more expensive original material? Are there hidden flow-control restrictions which drove the original design? Untapped and undocumented knowledge usually resides with the system or equipment operator. With their knowledge of and input into the use of the part, the reverse engineering team can minimize the assumption-based decisions made and lower the risk of failure. In a case such as this, the user's assistance would be needed to investigate the original intent behind this design choice. Perhaps there is a design constraint on the system which drove the original requirement; or perhaps the original manufacturer simply told the shop personnel 20 years ago to "just do it" and the creative machinist did the best job with the requirement given and drilled, formed, and welded the final product. It has happened both ways. Hidden system requirements may not be obvious to the reverse engineering team but may be apparent to the user. By communicating effectively with each other, the reverse engineering team and the system or end user can avoid spending time in unproductive directions searching for technical answers where none exist.

Project Tracking

There is a need to track a potential reverse engineering project from its origins as a candidate in the prescreen to its final outcome, whether that means going the four full stages of reverse engineering or terminating anywhere along the course. The most effective way to track items considered for reverse engineering is to assign each item, component, or assembly a separate and specific number which can be used as a singular reference as the unit progresses through each stage of the entire reverse engineering program. This reference number should not be the manufacturer's part number because many parts have similar numbers. The part numbering schemes which manufacturers use to identify individual products vary widely; therefore, in a large reverse engineering program different parts run the risk of having identical part numbers. In a large reverse engineering program there may be 6 valves or 10 different circuit cards being evaluated at any given time. Confusion would quickly set in if individual project numbers were not assigned. This assigned project reference number will correlate to, but should not match, the manufacturer's name and part number. This project-specific reference number will be needed for *all* reports generated in the reverse engineering process. It is also very useful for the purpose of tracking project costs.

To make a reverse engineering program really work for the benefit of the organization which has chosen to implement reverse engineering, a cost-tracking system for each project must be in place. To meet an objective means there must be a way to measure it. The most effective measurement system for reverse engineering projects is accurate economic predictions versus final project costs. The economics of a project revolve around the return on investment which is equivalent to significant cost savings over time. Overall program success can be gauged by the average return on investment for a series of components and their success in increasing the overall reliability of the system. The project cost:savings ratio is the ideal index. Project costs must be measured from stage 1 through stage 4 to assess the true cost to reverse engineer. The prescreening cost is significant but should be considered to be a sunk cost as it is a calculated investment. For this investment the plant manager will be able to intelligently gauge where the true problem components lie, and that is a valuable piece of knowledge, one well worth the prescreening investment.

Some people consider the cost invested to accomplish thorough prescreening to be excessive. Often hundreds of candidates must be reviewed to find only one-tenth of these candidates can be considered for review in stage 1. Some candidates never pass from stage 1, while others fail to finish the full four stages of reverse engineering. In a reverse engineering program which failed, over $6 million was invested into the prescreening process to review over 11,000 parts. This review netted about a dozen good candidates before those overseeing the review became frustrated with the small turnout. This frustration led to poor candidate selection, with an increase in borderline candidates. One single project alone cost over $500,000, of which $250,000 was a cost overrun on a part, which in the end would have cost at least twice as much after reverse engineering as it would have to continue buying from the available source. After much bickering and finger pointing, the program folded, leaving the reverse engineering team over $250,000 in debt. In retrospect this can all be traced back to poor candidate selection, as well as some very poor communication and project oversight. If the cost of selecting good candidates is too high, it will only be more expensive to quit while you are behind. Invest in projects wisely, track costs conscientiously, but above all, prescreen candidate selections carefully.

A specific charge account should be established for each project to accomplish this tracking. The final results and economic viability will rest on the differences between projected and actual costs. The ability to measure the success of each project is paramount to the success of the program. Since success is measured by the return on investment, a return of 200:1 is very impressive only if the numbers to back up this claim reside in an engineering or corporate database.

The Reverse Engineering Database

Although a cost-tracking system is not a requirement, particularly if only a few projects are being considered, it is recommended for more extensive programs with many candidates. This need to track project costs also brings to light the need for a unified database to track a project from inception through implementation. Ideally the database should be able to track all the following characteristics:

Item name and all identifying information

Economic and logistics data

Projected return on investment

Project type

Data prescreened

Data sent to stage 1

Project cost estimates

Number of samples obtained

Location tracking for stages 1 through 4

Number of prototypes built

Final unit cost

Final project cost

Actual return on investment

Additional information can be added as necessary to measure and track the progress of parts through the reverse engineering cycle. Since there is a vast amount of data to be tracked and these data are stored in the form of dates, costs, and whole numbers, the selection of a program to track the data is important. The database should be able to track many candidates and their final fates as well as many types of data. The type of database selected should allow for the data to be tailored for reports on progress to management or the customer. A unified database that can handle all the information and produce reports is recommended to avoid duplication of information in many small databases.

Candidate Prescreening

A product-line manager would generally not be considered fortunate to have chronic manufacturing process failures . . . unless these faulty components were in the candidate pool for reverse engineering. The larger the pool of potential reverse engineering projects, the better the chance that reverse engineering can improve productivity. According to this logic, the worst operational capability in a production facility is the best source of potential reverse engineering candidates. What is being examined is an underfunctioning system as a whole. Emphasis is on the whole system. It is wise to always view the component within the realm of the system. In particular, those items which are lowering the operational productivity or effectiveness are being targeted for improvement through reverse engineering. Prescreening is the method used to conduct candidate assessment, analysis, and selection. Figure 3-1 gives an overview of the prescreening process.

The components in a targeted pool of problem components are known as *candidates*. These items have not yet been fully prescreened. A candidate can be either a singular item, part, component, unit, or subassembly and may contain many smaller parts, but it is either purchased, sold, marketed, or otherwise described as a single entity. Referring back to Fig. 1-2, illustrating how parts develop into programs, a *candidate* that has not been prescreened cannot yet be considered a *project*. The transition from candidacy to project is not made simply by prescreening but also by acceptance into stage 1 for full evaluation and verification. With its induction into the four-stage process it is then considered to be a project. Candidates can be prescreened and considered to be poor candidates (ones with a low priority or small chance for success), in which case they will simply fall into a category similar to "presidential hopefuls." If the time and money are available, they may be considered at a later date. Prescreening is the part of the reverse

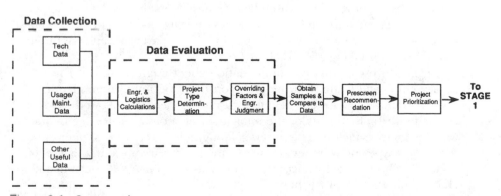

Figure 3-1. Overview of prescreen process.

engineering process, which focuses on the selection of those items which can increase the effectiveness or productivity of the system being reviewed.

The Qualities of a Good Candidate

The following questions now surface: How are good candidates chosen? What are the selection criteria? A high likelihood of success is needed before investing time and effort into each project. Remembering that success is measured largely by return on investment, the bottom line cannot be forgotten altogether. Reverse engineering still is a business venture and an investment in the future of a system or production capability. A good candidate often exhibits a high failure rate, high annual usage, or simply costs too much. It is not so technically complex that the cost to reverse engineer would be prohibitive. The part being reviewed should not be so critical to the operation of a system that catastrophic failure of the component would cause loss of life or destruction of the operating system. Technical data availability and adequacy must also be addressed. An exceptional candidate has a combination of both economic and logistics factors in its favor. Other overriding factors such as lack of supply support, obsolescence, or patent rights must be accounted for when deciding on a candidate's viability for the full four stages of reverse engineering. For the most part the qualities of a good candidate will be evidenced in the following characteristics (not in order of priority):

- Economics
- Logistics
- Return on investment
- Technical complexity and criticality

Actual prescreening of the candidates is based on the analysis of paper (projected, estimated) data. The physical part is not used in the prescreening process because simple visual inspection of the part while analysis of the available data is taking place is of little use.

A typical prescreening recommendation sheet might look like Fig. 3-2. Prescreening will begin with the inclusion of all pertinent item identification. Candidate assessment begins with data collection followed by data evaluation, which consists of the analysis of economics factors combined with the logistics factors. The reviewer's judgment is required to determine the project type and whether any overriding factors are present, the level of relative technical complexity, and overall assessment of the candidate's chances for success. A further recommendation to proceed to stage 1 requires the determination of the number of operational and nonopera-

---

RPT#_____
DATE_____

PRESCREEN RECOMMENDATION SHEET

P/N_____   Other Identifiers:_____
Item Name_____
Mfgr_____
        Mechanical _____   Electrical _____   Other_____

TECHNICAL DATA AVAILABILITY
Detailed Drawings   Rev. On Hand   Latest Rev   Restrictions?

Technical Manuals   Performance Specifications

Summary of Data Inadequacies:

        ECONOMIC FACTORS   LOGISTICS FACTORS

Unit Cost:_____   Annual Usage:_____
Target Unit Cost:_____   Remaining Service Life:_____
Annual Cost:_____   Available Assets:_____
Life Cycle Cost:_____   Life Cycle Usage:_____
Life Cycle Savings(LCS):_____   Part Population:_____
Projected Cost to RE:_____   Replacement Rate:_____

Projected Return on Investment:_____
        ROI= LCS - RE Cost
              RE Cost

DETERMINATION OF PROJECT TYPE:
Product Verification_____   Data Enhancement_____   Data Development_____

OVERRIDING FACTORS:

TECHNICAL COMPLEXITY:_____   (5=COMPLEX : 1=SIMPLE)
ENGINEERING JUDGMENT:_____   (5=EXCELLENT : 1=POOR)

FURTHER RECOMMENDATIONS:
Proceed to Stage 1: YES____   NO____
Number of Samples Needed:_____ Operational   _____ Non-operational
Other:

PROJECT PRIORITY:

Engineering Staff Signature

Figure 3-2. Typical prescreen recommendation sheet.

tional samples necessary. Further recommendations may not be reverse engineering at all but could involve a search for a commercially comparable product or to purchase additional data from the original manufacturer to complete the technical data package. Project prioritization can occur at this point or be assigned after a group of candidates are prescreened.

The prescreen recommendation sheet is simply a typical recommendation sheet and it both can, and should, be altered to suit the needs of a particular organization conducting reverse engineering. The information shown is the minimum needed to assess the candidate. If additional information is needed, simply expand the realm of the prescreen. Only by experience can iterations of reverse engineering be perfected to address unique needs.

Data Collection

Before the prescreen recommendation sheet is completed, all available data must be collected. Data collection may be as arduous as any research task, but thorough data collection will have the effect of reducing the overall project cost. Drawings and technical manuals should be included. Usage and maintenance data is needed along with performance specifications. Any available operational parameters are helpful while not necessary.

Technical Data and Manuals. Useful technical data includes all drawings and technical manuals. Engineering drawings include layout, detail, assembly, installation, arrangement, and control drawings and schematics and wiring diagrams. A complete listing of drawing types can be found in American Society of Mechanical Engineers (ASME) standard Y14.24M— 1989, *Types and Applications of Engineering Drawings*. (See App. A.) This document covers engineering drawing and related documentation practices and has been coordinated with both the International Standards Organization (ISO) community and the United States Department of Defense (DoD), which, as major participants in the standards world, are also trying to establish a common practice and understanding of the various types of engineering drawings. The use of standard drawing types and their application will assist in technical data interpretation to produce a consistent final product.

All applicable detailed and higher-level assembly drawings must be collected or identified as available, current, or missing on the prescreen recommendation sheet. Parts lists will need to be included. Any technical manuals defining the operating parameters of the system or higher assembly are also sought. The following questions need to be answered.

Are detailed drawings available?

Are there any higher-level assembly drawings?

Are there multiple drawings for this component?

Are any of these additional drawings missing?

Are these drawings complete?

What is the latest revision of each drawing?

Is the available drawing the latest revision?

Is any of the data proprietary, restricted, or patented?

Is there any additional information available in technical manuals?

Any drawing or technical data that are restricted, such as a proprietary or limited-rights drawing, can be reviewed in the prescreening stage but cannot be used in any of the reverse engineering stages. Restricted data cannot be passed beyond the prescreen stage. Noting that restricted data exist and their relative level of completeness is sufficient, but the data must remain unavailable for the reverse engineering stages 1 through 4.

Generally, extensive data will not be available for a patented part, and any patent identification that is available will not be evident until the actual part is visually inspected. If the part is only one component of an entire assembly, reverse engineering can continue using the patented part, but if the entire assembly is patented, prescreening of this part should not continue and the part should not be reverse engineered.

The summary of data inadequacies serves to highlight in a general sense which pieces of data are missing and the level of completeness of the available data. At this juncture in the reverse engineering process, it should become evident which data are missing and need to be developed.

Usage and Maintenance Data. Usage data concern the number of parts used on an annual basis for both component failure and routine maintenance. *Usage data,* which are needed to conduct the prescreen analysis, consist of unit cost, number of units ordered per year or the number of parts used per year, the number on hand in the supply system, and the number on back order. The part population is determined by the total number of applications of this specific part in both the system being examined and any other similar systems throughout the plant.

Usage data are similar, but not necessarily equivalent to, failure data. Often items are replaced as a part of routine maintenance procedures which are meant to avoid equipment failure. *Routine replacement* does not mean unit failure. A calculated replacement rate should reveal approximately how many units are being used per year. This annual usage includes both failure and maintenance requirements. Maintenance requirements include both preventive and corrective maintenance.

Consider a hypothetical situation where 100 units operate (part population = 100) and 50 of these are replaced annually; this indicates an abnormally high usage rate of 50 percent. If the failure rate is greater than 10 to 15 percent of the part population for electrical parts, this is considered to be high, and a corresponding failure rate of greater than 5 to 10 percent for mechanical parts is considered to be high. If the calculated *component* usage rate seems higher than the average *system usage* rate (if this number is known), this component's viability for reverse engineering candidacy may increase. As an example, if a system functions operationally 95 percent of the time and one unit fails 50 percent of the time, that unit's usage rate (of 50 percent) is 10 times higher than the average system usage rate (of 5 percent).

Other Useful Data.   Performance specifications are to be included in this section along with any reliability, availability, and maintainability (RAM) data. RAM data includes troubleshooting, test, inspection, repair, overhaul, and calibration data. Procedures for the conduct of any testing, inspection, installation, or operations are sought. Mean time between failure (MTBF) may be a form of logistics data, but, since annual usage rates are more important, it is not necessary for logistics calculations and can be introduced as other useful data. Average downtime might be another good piece of information to have for later calculations in the Engineering and Economic Summary report produced in stage 4 in the event the average downtime is improved as an added improvement to the system due to reverse engineering. Additional specifications regarding material composition, quality assurance, testing, or inspection methods should be included in the prescreen review since they will be necessary for testing any prototypes built in stage 3 which validates the reverse engineered design.

Data Evaluation

The raw data assembled in the data collection must now be processed in order to make intelligent decisions concerning the fate of this candidate.

Economic and Logistics Calculations.   Certain necessary calculations are made to assess the value of the numbers collected thus far. The following definitions are useful. Items are defined in the order they appear on the prescreen sheet.

unit cost: The current price for this item. While this may sound self-explanatory, unit pricing varies widely according to the quantity purchased over time. Usually a lower unit price can be achieved for buys (orders) larger than for smaller orders, which must be expedited to fill an urgent need. A unit purchased in advance might cost only $300 but might cost $800 or more if purchased in a last-minute panic buy. For the purposes of reverse engineering, any pricing that is a small lot or panic buy is considered to be an inaccurate reflection of the true unit cost. Pricing data for a medium-size buy prior to an urgent need is a truer basis for all reverse engineering calculations. Calculations based on an inflated price may result in reverse engineering and success based on a 25 percent decrease in unit cost; however, one may find that the original manufacturer is now offering the same item for 50 percent less than the panic buy price, and all will have been for naught.

target unit cost: The projected cost to obtain the part after reverse engineering. The goal here is a 25 percent decrease in the unit price. This is a minimum goal used only to protect the program from too many projects with slim returns on investment. It contains a margin for error as a form of safety factor to protect the reverse engineering team and the company's best interests. It is not meant to be a rigid guideline, just a rule of thumb in the prescreen process.

annual usage: The number of parts used per year for both maintenance and failure requirements. It should be averaged over as many years worth of data as can be found. If the usage rates during the first years of system implementation were much lower than current levels, it may be reasonable to eliminate very early data. Few parts should be replaced in the early stages of system implementation, but as the system matures this usage rate should normalize. Taking the year of highest usage can also present skewed results with respect to later calculations such as return on investment. Ideally, a normalized annual usage would be helpful for economic calculations, since early usage rates will be lower than usage rates experienced toward the end of the system life cycle.

annual cost: The unit price multiplied times the annual usage rate. This equals the monetary outlay for the procurement of this component for one year, preferably one good representative year based on normalized usage, not the year of highest usage.

remaining service life: The remaining service life of the operating equipment can be determined by approximating the length of time until a planned replacement system is to be installed. A 30-year service life for equipment is not uncommon for mechanical systems, and approximately 10 to 15 years is a good estimate for the length of the service life for electrical and electronic systems. Most mechanical systems begin to deteriorate after only about 10 years of operation, while electrical and electronic systems can begin to deteriorate after only 5 years of operation.

available assets: The total number of parts in inventory at the time these prescreen calculations are made. Back-order parts, or those orders already placed but not delivered by the manufacturer or supplier, can also be accounted for in available assets since these parts will probably be on hand before reverse engineering can be completed or the purchase of which cannot be cancelled.

life-cycle usage: The annual usage multiplied by the projected remaining service life minus the available assets, including those on back order. This represents the total number of parts which must be obtained to maintain the system through its lifetime.

part population: The total number of parts in the entire system, not merely the singular *subsystem* being studied for reverse engineering. This accounts for all additional applications of this part within the extra production facilities. Since reverse engineering a part may have additional applications in other areas, this total part population will point to other subsystems which could benefit from the reverse engineering process.

replacement rate (also failure rate): The ratio of the annual usage divided by the part population times 100 to express it as a percentage. As pointed out earlier in this chapter in the section on usage and maintenance data, this indicator only serves to either add to or detract from this item's reverse engineering potential. The replacement rate is part of the engineer's judgment and later becomes embedded in the quality evaluation report (QER). It serves as an indicator of a reliable component or a source of frequent trouble if this component replacement rate exceeds the system usage rate. (See page 48.)

life-cycle cost (LCC): Equal to the life-cycle usage multiplied by the unit cost or the annual cost multiplied by remaining service life if there are no assets on back order. Either calculation should net the same approximate result.

life-cycle savings (LCS): Computed by multiplying the LCC by 25 percent. The 25 percent represents the 25 percent target decrease in unit cost. This equals the predetermined decrease of the procurement cost, which is the desired result by reverse engineering. If reverse engineering is to be conducted, as a target, 25 percent of LCC should be saved. When the cost to reverse engineer is subtracted from the LCS, a positive net LCS should result. The net savings is the true value on which the return on investment will be based. In this manner, there is some very real return on investment despite the risk and the cost to do the actual reverse engineering. LCS is needed for the return-on-investment calculation.

projected cost to reverse engineer (RE cost): An estimate, best guess, often costing many thousands of dollars based loosely on the technical complexity of the part and the amount of available data. An estimate will suffice for a ballpark figure on the prescreen recommendation sheet. A guesstimate of 10 to 100 times the unit cost may be used unless better data are available through either experience or insight. The stage 1 project cost estimate will reflect a more accurate cost using a detailed breakdown of the steps needed to complete the project. The prescreen estimate is necessary to make an initial recommendation. In calculations, this quantity is shown as "RE cost" whether it is (1) projected cost to reverse engineer from the prescreen, (2) estimated cost to reverse engineer for stages 1 through 4 from the stage 1 report, or (3) actual cost to reverse engineer calculated in the final Engineering and Economic Report in stage 4.

return on investment (ROI): The ratio of the projected LCS minus the projected cost to reverse engineer divided by the projected cost to reverse engineer. This results in the ratio of the net savings to the reverse engineering cost and is the cost-effective measure of reverse engineering an item. It is calculated by subtracting the cost to reverse engineer from the projected LCS divided by the cost to reverse engineer.

Return on Investment. Some simple ROI formulas follow.

$$ROI = \frac{LCS - RE \ cost}{RE \ cost}$$

or

$$ROI = \frac{net \ savings}{RE \ cost}$$

Most calculations will be simply expressed as

$$ROI = \frac{LCS - RE \ cost}{RE \ cost}$$

Some simplified hypothetical examples will help to lead you through many of these calculations.

Example. The whatzit costs $300. The whatzit is part of a system which needs to be maintained for only 10 more years. Routinely 500 parts are used per year. There are 2000 parts on hand with 1000 on back order.

The following economic information can be determined. The 2000 parts in stock will maintain the system for 4 years at 500 parts per year. With 1000 on back order there is 2 more years of supply. This means that 6 of the remaining 10 years worth of repair parts are available and only 4 years remain in the life cycle for which there are no parts.

The 4 remaining years will require approximately 2000 parts at a usage rate of 500 parts per year. If there are no shelf-life restrictions, these 2000 parts can now be purchased for an investment of $600,000 ($300 per part × 2000 parts). A few considerations are in order before reaching a conclusion to purchase the remaining 4 years of supply now.

If $300 per whatzit is considered to be a reasonable cost, it makes good economic sense to purchase only 4 years of supply now or later on rather than invest in reverse engineering.

If, by reverse engineering, the whatzit unit price can be reduced by 25 percent (the target unit price reduction) to $225 each, then the 2000 remaining parts can be purchased for $450,000 ($225 × 2000). This would result in a $150,000 LCS over the remaining life of the part. LCS can also be calculated by multiplying LCC ($600,000) by 25 percent (0.25), which also accounts for the 25 percent reduction in the unit price. $600,000 × 0.25 = $150,000.

From the $150,000 LCS the cost to reverse engineer would need to be subtracted out. If the cost to reverse engineer is $30,000 (assume the reverse engineering cost to be 100 times the cost of the item) the return on investment can be calculated also:

$$\text{ROI} = \frac{\text{LCS} - \text{RE cost}}{\text{RE cost}}$$

$$\text{ROI} = \frac{\$150,000 - \$30,000}{\$30,000} = \frac{\$120,000}{\$30,000} = 4{:}1$$

Return on investment is expressed as a ratio. If you recall the guidelines from Chap. 1 (section on the risk of failure) that prescreened components should not enter stage 1 without a projected ROI of 25:1, then the reverse engineering of the whatzit with a projected ROI of 4:1 presents a risk due to the slim margin for return. At this point it is simply the engineer's and management's judgment call on proceeding further into stage 1.

Example.   The whatzit again costs $300 per part. The system has 22 years of remaining service life. There are only 1000 parts on hand. Annually, 500 parts are replaced because of system usage. Is there a significant return on investment sufficient to justify the potential of reverse engineering the whatzit?

Using 500 parts per year, there is a 2-year supply available. There are 20 (22 − 2) years left to maintain the system. The 20 years × 500 parts per year = 10,000 parts to be supplied over the remaining service life.

LCC = 10,000 parts × $300/part = $3,000,000. LCS = $3,000,000 × 25 percent (0.25) = $750,000. The reverse engineering costs again are estimated to be $30,000.

$$\text{ROI} = \frac{\text{LCS} - \text{RE cost}}{\text{RE cost}} = \frac{\$750,000 - 30,000}{\$30,000} = \frac{\$720,000}{\$30,000} = 24{:}1$$

A projected ROI of 24:1 is worth pursuing when the net savings is as large as $720,000. This leaves plenty of room for error and still points the project to stage 1 for a full and complete evaluation of additional factors which cannot be accounted for in the paper review of the prescreen.

The Go/No-Go Matrix: A Reverse Engineering Consideration.   Economics is the most common driving factor in the decision to reverse engineer an item. This is reflected in the projected return on investment. If a few thousand dollars can be invested to solve a nagging operational defect, the decrease in the repair costs compounded with a decrease in replacement parts costs make reverse engineering a justifiable capital expense. The decrease in system maintenance costs by lowering the system downtime using a more reliable part, or the decrease in repair costs, is not accounted for in the prescreening but is calculated in stage 4, if those figures can be calculated for the Engineering and Economic Summary Report.

If an item is relatively technically complex, the reverse engineering cost will increase. The increase in project costs due to increasing levels of technical complexity is not always linear; in many cases the increase due to technical complexity is exponential. An increase in the reverse engineering costs corresponds to an increase in the risk involved. Essentially, the prescreening process is about managing the amount of risk inherent in the overall program. Figure 3-3 plots the relationship between cost, complexity, and risk schematically. This scheme does not account for all factors necessary for a good potential reverse engineering candidate, but shows a common and basic set of relationships.

Determination of Project Type. The determination of project type is part of the prescreening process which accounts for both the type and

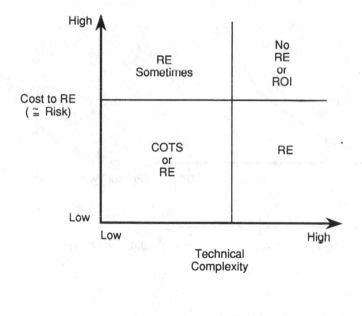

RE = Reverse Engineering

ROI = Return on Investment

COTS = Commercial Off-The-Shelf Item

Figure 3-3. Go/no-go matrix (RE = reverse engineering; ROI = return on investment; COTS = commercial off-the-shelf-item).

amount of data available. The available data directly influences the cost and risk factors. Figure 3-4 shows this relationship graphically. Again, this only illustrates a basic trend not a full evaluation of all factors.

More data, less risk, and lower cost are good signs pointing toward the use of reverse engineering. When the technical data appears complete but there is some doubt as to its usefulness, there is product verification. If there is some technical data available, the project requires data enhancement. Data enhancement is performed whenever there is any amount of missing data. If there is no technical data available, the project requires full data development.

Product Verification.   The least expensive and least risky reverse engineering project type is product verification. Product verification can resolve

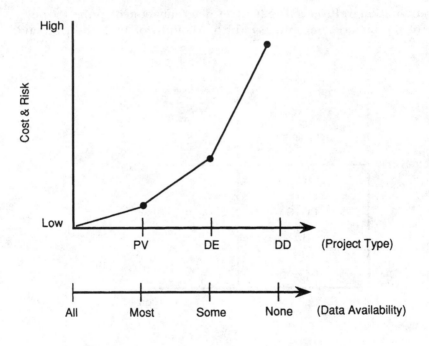

Project Type ≅ Technical Complexity

PV = Product Verification

DE = Data Enhancement

DD = Data Development

Figure 3-4. Relationship of cost and risk to data availability (PV = project verification; DE = data enhancement; DD = data development).

suspicions about completeness or verify hunches that there is more to a project than could be determined during the prescreening process of reverse engineering. On the surface, it appears that all the available data is complete and accurate but current drawing revision levels are unknown or material verification must be performed. In product verification usually only one sample is needed to compare the part to the detailed drawing on hand.

Data Enhancement. The typical reverse engineering project falls into the data enhancement category. A detailed drawing could contain every dimension necessary to manufacture a duplicate part with the exception of the tolerances, a material specification, or a schematic. Data enhancement might also be needed if there is only an outline drawing or a performance specification available. Depending on the amount and completeness of the technical data, this type of project usually requires three to five sample parts (both operational and non-operational) to increase the chances for success during reverse engineering.

Data Development. Data development is the most difficult project type. To limit the amount of risk, 5 to 10 sample parts will be needed (when they can be obtained) to accurately develop a technical data package that reflects the part. Data development can be a simple matter for a simple part. It can become a nightmare when material compositions are not easy to determine, tolerances have to be guessed at, or the item provides a nasty surprise by possessing an internal cavity filled with an unknown lubricant or a circuit card which has three extra layers not previously mentioned. As a precautionary note, both the risk and the cost of this project type are highest. The return on investment could exceed 500:1, or the failure could cost thousands or even hundreds of thousands of dollars.

Overriding Factors. Two overriding factors can give a candidate high priority on the list of potential reverse engineering projects. In either case all the usual factors, such as technical complexity and economics, can be overridden by a higher need to maintain system operations at their current levels if the reverse engineering costs are not exorbitant or the criticality of the component warrants an "at all costs" attitude.

Obsolescence. Obsolescence is due to a decline in the need for a technical capability over time to the point where the component is no longer available through any supply system. Obsolete components sometimes make excellent reverse engineering or value engineering candidates. Obsolescence may trigger the need to upgrade an entire portion of a manufacturing capability, or a small batch of reverse engineered parts may keep a 40-year-old production facility operational until the company can acquire the funds necessary to reinvest in equipment upgrades or system modernization. As mentioned earlier, an obsolete component for an existing system is every line manager's nightmare. The show must go on, yet no one makes replacement parts any longer. This is a perfect candidate for reverse engineering.

Lack of Supply Support.   A lack of supply support is closely related to the problem of obsolescence, but it is not quite the same. Original manufacturers who have gone out of business are nothing new, but the current pace of technological progress is changing so rapidly that there is a risk of losing suppliers as well as the current sole manufacturer. Often suppliers are not available because of low demand for a part, and it becomes uneconomical for the supplier to stock this item. It may be even more difficult to locate an alternate supplier elsewhere in the country or in the world because of language differences or trade restrictions. This does not truly indicate obsolescence because it is being manufactured somewhere, but that actual place is indeterminate from where your needs stand.

Component Criticality.   The two overriding factors described above can forward a project into reverse engineering. Only component criticality can terminate a project despite all other data indicating continued progress. The technical complexity of a component is not equivalent to component criticality. If the component is part of a system, such as a Level 1 SubSafe on submarines, which is designated to be of such high integrity that it should not be considered, the potential for gain through reverse engineering would be negated. The loss of life or potential for catastrophic failure of the part within the system may make this part too critical to take the risks inherent in reverse engineering. To not consider the possibility of an overly critical part in prescreening could mean that the part would be terminated further into the reverse engineering process and render futile all work to improve this part. If a part is essential to the operation of a critical system, do not under any circumstance reverse engineer this component. A critical component should not even be considered a candidate, except under unique conditions determined by end users or management.

Engineering Judgment.   The technical complexity and the overall assessment are simply the reviewing engineer's best judgment based on all the available technical, economic, and logistics information. Prescreening is based on an evaluation of the physical data available. Most often this includes paper drawings, performance specifications, or failure data based on the number of replacement parts ordered over a period of time or the number of hours spent in downtime based on equipment failure. The physical part is not in hand as yet. Again, it is the paper evaluation based on the best judgment of the engineer who is responsible for reviewing the documentation and performing the prescreening. One of the last items on the prescreening document is the judgment factor on a scale of 1 to 5. Here the engineer can express an overall assessment of the probability of success this item can have in reverse engineering. It can also be used by upper management to prioritize which projects are funded first.

Engineering judgment is used throughout the prescreening and the four stages of reverse engineering in many situations. It is paramount to keep in mind that the use of good judgment is an exercise in personal responsibility. Many recent catastrophic engineering failures have been traced back more to poor judgment than to technical negligence. Exercising your design responsibilities is not a luxury; it is a paramount requirement.

Determination of Samples Necessary

As discussed in the section on determination of project type, above, project type determination influences the number of samples needed to conduct reverse engineering. In most cases fully operational sample parts are needed but sometimes these are not available and failed or nonoperational parts will have to be used as a baseline. Although non-operational samples increase the risk of undertaking reverse engineering, they can be useful later in stage 1 for fault isolation or failure analysis. As the number of samples increases, so does the variation in the physical dimensions, and wide dimensional swings have been noted on parts which have drawings that specify very strict tolerances. This can lead to the conclusion that indeed no tight tolerance requirements exist and the manufacturer could be buffering a bloated profit margin on a necessary replacement part. Another reason for wide dimensional variation could be that the manufacturer is actually buying from multiple suppliers with widely varying quality assurance practices or any number of other poor production practices.

Visual ("Eyeball") Inspection.   There is an interim phase after the engineer has found an excellent candidate and management has funded this project when the part has been not only ordered by also delivered. Once delivered, the part is inspected by the engineer who completed the prescreening. A routine visual (also referred to as "eyeball") inspection will verify that the physical (actual) part matches the paper (specified) part. This may sound trivial, but a small percentage of parts are incorrectly labeled or stocked. This seemingly trivial logistics problem could actually be the cause of the high usage rate of the part. The equipment fails, the repair technician isolates the faulty part, the maintenance department turns in a requisition for the specific part number, the purchasing department orders the part now specified by a part number, shipping forwards the requisitioned part to the repair technician, who cannot repair the equipment because it is still the wrong part, and the cycle continues.

Anyone who has tried to do home auto repair and gone into an auto parts store armed with the correct part number, type, and size (plus the year, make, and model of the car) can attest to the problems encountered when the original manufacturer of that part has replaced it with a newer model that has a three-wire hookup instead of the old two-wire configuration, or the manufacturer's catalog shows that no such part exists and the helpful clerk suggests that this alternative part is what always works for the rest of the entire world. On bringing such a part home it is physically impossible to put this new and improved part into the old part's former home. The same difficulties can occur on a grander scale with major systems. This makes the seemingly trivial act of a quick eyeball inspection of the part, preferably by the same person who conducted the paper analysis, necessary to the overall success of a reverse engineering program.

Project Prioritization

If a group of candidates are prescreened at one time, it is wise to group, or rank, them according to the likelihood of success or criticality of need. The criteria for grouping or ranking are based on the highest needs of the system being evaluated for reverse engineering. If obsolescence is the overriding need, then the obsolete items have a higher priority over those being considered on a cost savings basis. Likewise, if the overriding goal is cost savings, then those with the highest potential for cost savings are considered first. Either way, the system needs determine the reasons for prioritization.

The use of simple classification categories is recommended. Three broad categories should suffice, such as high, medium, or low. Other choices might be obsolescence, cost savings, or simplicity. The simpler the category, the better at this point in the evaluation phase. Overcomplicating the ranking process is a waste of energy and resources and will not add to the evaluation process.

On to Stage 1

Let us assume that this candidate has passed a prescreening and still looks like a good reverse engineering project, on paper. One final task must be done before handing the potential project off to the core reverse engineering team: assembling all the data, both collected and evaluated, less the restricted information, and passing this information on. A copy of the prescreen sheet must accompany the technical data. All the unused and failed samples must be forwarded. In short, make a copy of pertinent information and note what date the sample parts, technical data, and prescreen evaluation is handed off for a stage 1 review.

4
Stage 1: Evaluation and Verification

The candidate pool has been searched. The components have been ranked by priority. We will begin with the component selected with the highest expected return on investment, the highest likelihood of success, or the highest priority. All available technical data has been gathered and evaluated. Missing data and project type have been identified. Sample parts, both unused and failed, have been obtained. All the preliminary prescreen work has been completed, the groundwork has been laid, the multidisciplinary team will soon be assembled, and the four stages of reverse engineering can begin.

Stage 1 is the most demanding part of the reverse engineering process. More actions must be taken in stage 1 than in any other portion of the reverse engineering process (although technical data generation, stage 2, or prototype testing and evaluation, conducted in stage 3, may be more engineering intensive). Stage 1 requires multidisciplinary teams to work independently, yet to rely on each other to determine what information is essential to continue the process and what ancillary information may be needed later in stages 2, 3, or 4. Figure 4-1 provides an overview of stage 1.

The primary steps in stage 1 are as follows.

- Visual and dimensional inspection
- Discrepancy review versus available data
- Failure analysis

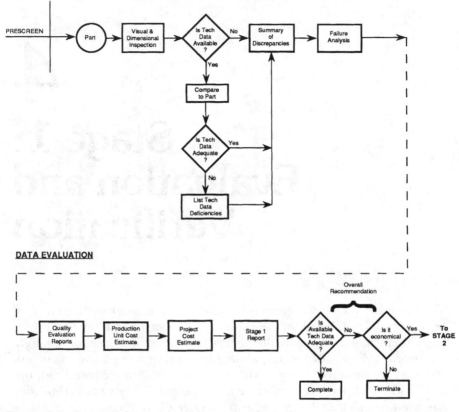

Figure 4-1. Overview of stage 1 (evaluation and verification).

- Quality evaluation report
- Stage 1 report
- Go/no-go decision

The visual and dimensional inspection, discrepancy review versus available data, and failure analysis constitute the *data collection* portion of stage 1. The quality evaluation report, stage 1 report, and go/no-go decision constitute the *data evaluation* portion of stage 1.

Stage 1 entails the complete characterization of a part using visual and dimensional inspection, material analysis, and identification. Comparisons to available data must be made. A failure analysis is conducted if failed sample parts have been obtained. Then, a quality evaluation report (QER) is generated. A stage 1 report must also be generated complete with the projected reverse engineering cost estimates. A final go/no-go decision must be reached by both the project leader and the approval body on the basis of the available information.

Complete records of all tasks completed in stage 1 should be kept in a project file with the prescreen project reporting number listed on all records. (See Chap. 3, section on project tracking.) The data collected in stage 1 are used to generate the QER and the stage 1 report. (See sections on QER and stage 1 report later in this chapter for a delineation of what information collected in stage 1 must be included in these reports.) As in the prescreen, the information is first collected and later evaluated in report format.

Reverse Engineering Team Meeting

After a preliminary study of the part the lead engineer determines which basic technical specialties, such as drafting, welding, or machining, are needed and then selects the reverse engineering team accordingly. It is particularly expedient to hold a formal meeting to launch each reverse engineering project. The lead engineer collects all the data and samples in one location and invites the project team members to participate in a collaborative meeting.

Later, in stage 3, the technical data package and testing of prototypes in an operating system will require the approval of system supervisors, production managers, or others responsible for technical data package oversight. It is good protocol to invite these individuals to the preliminary reverse engineering team meeting to assure them that this part will not adversely affect the functioning of "their" system, and later give them some confidence that their needs have been included. This will help gain their approval and expedite any testing requirements if their endorsement of the project has been gained up front.

The preliminary reverse engineering is somewhat similar to an NFL (National Football League) team meeting in the locker room prior to the Sunday morning kickoff (or the Saturday morning kickoff for you college football fans). The head coach (lead engineer) lays out the game plan after studying all the strengths of the home team (end product of process) and the weaknesses of the other team (the sample parts and data, for example). The head coach, who will not actually participate in the game, will call the plays from the sidelines. The goal, of course, is not to score points but to make another component that will be at least as effective as the original one. This meeting can serve many purposes, but its primary objective is to set the direction of the game (project) and a tone of cooperation among team members. After a few projects the head coach (lead engineer) may find that only the primary players (who may vary from project to project) need to meet since the ancillary partners (who tend to remain the same) will already understand their role.

A copy of the prescreen data sheet along with all sample parts should be forwarded with *all* the technical data collected and evaluated from the pre-screen research. The only data that should not be forwarded to stage 1 are proprietary or restricted data. Drawings with limited rights should have been carefully reviewed during prescreen before being forwarded for inclusion in the stage 1 raw data package. Having all this information available at the team meeting is critical in order to make decisions regarding the direction this project will take. Missing data in the team meeting can produce erroneous results later in the process and may add unnecessary additional expenses to the project.

Key design features to look for in evaluating any design are a combination of any of the following:

Strength	Reliability	Wear
Corrosion	Friction	Processing
Utility	Cost	Safety
Weight	Noise	Styling
Shape	Flexibility	Size
Surface finish	Control	Stiffness
Lubrication	Maintenance	Volume
Stable artwork (for electronics)	Specifications	Obsolete parts

Since every component is unique, this is the time to consider all the tasks to be accomplished and the design options and pitfalls, to divide up the work needed to complete this project, and, perhaps, to lay out a timeline of which tasks can be done in parallel and which must be accomplished in series. Assigning the tasks needed in each stage of reverse engineering will optimize the time it takes to complete a project. A discussion of the workload each team member anticipates can indicate areas which can become bottlenecks prior to completion. A full and open discussion of the objectives, such as whether this project is a simple product verification or a complex data development situation, can avoid situations of overkill by one party in their function. This may be a time to discuss whether value engineering can be considered, but perhaps that can be determined only later in the process. The team leader sets the tone, pace, and direction, and all parties have the option to review their roles, particularly early in a reverse engineering program.

When all the project engineering team members leave the meeting they should fully understand their roles and responsibilities. If there is overlap of duties between members, these parties can discuss and try to resolve any areas of conflict. Overlap may occur when one individual has more than one function to fulfill, as, for example, duties of the draftsperson and the drawing checker, or of the inspection and test personnel.

If later in the process one person, say, person A, leaves the project and is replaced by person B, another individual of like background, the lead engineer should not have to coach this newcomer. Person A should be able to review with person B the details covered in this reverse engineering team meeting. This can prevent confusion and redundant sets of directions, as well as eliminate the need for meeting after meeting after meeting. Unlike football, you do not want to have a huddle after each play.

Visual and Dimensional Inspection

The visual and dimensional inspections constitute the first task conducted in stage 1. The *visual inspection* is simply a review of the overall condition of the part in terms of reproducibility, quality, and its present state of deterioration or wear. (Using appropriate engineering judgment becomes a challenge when the failed part is in a state of grave deterioration.) The visual inspection conducted in stage 1 is similar to the eyeball inspection in the prescreen except that the goal of the visual inspection in stage 1 is not to determine conformity to the paper data but to assess where to begin the process of reverse engineering. It includes the location of points of reference for machining, key design elements and materials, and notations on any special materials, coatings, tolerances, and so on.

The *dimensional inspection* is a complete and accurate measure of all component dimensions needed to fully characterize the sample part and establish a configuration baseline. The configuration of the part is defined by its size, shape, weight, tolerances, finishes, and other parameters. No wall thickness or internal diameter can be overlooked. The methods of joining inseparable parts must be inspected. The order of assembly must be discerned so that the sample part(s) can be disassembled without impairing the intent of the original design.

Dimensional inspection is conducted using calipers, micrometers, coordinate measuring machines, or any other tool of measure. Optical or laser tracking with probes which automatically generate a digitized computer model of the design are also very valuable in expediting the data evaluation and generation tasks. Computerized and digitized measurement systems are extremely helpful when components with moving parts are analyzed. An example of this computer-aided design–computer-aided manufacturing (CAD/CAM) utility would be during the dimensional analysis of a valve. An automated computer simulation of the movement of the internal parts is valuable when determining tolerances and clearance fits. Overall, a computer modeling capability will expedite the collection of measurement data and regeneration of the original design. Modeling is also very useful when upgrades or changes to the design become necessary after prototyping or during value engineering.

Tolerance determination is one of the most challenging parts of the dimensional inspection. In many cases the mating component(s) is (are) not always available at the time of the dimensional inspection. This will be a true source of turmoil for the design engineer who would rather not guess at such things. The original technical data may call for tolerances in the $\frac{1}{10,000}$ths but there may be dimensional variations in the sample part in the range of only one-tenth of an inch or centimeter. It shows no real design or common sense to have such tight tolerances when the component dimensions are so varied. If any one of the parts sampled is expected to replace the original, and a 0.1-in difference is measured on a design feature, there is no reason to machine the component to a $\frac{1}{10,000}$th tolerance. Engineers, with their inherent need for accuracy, find these situations beyond good reason and have been known to tear their hair out trying to make a final determination. Good engineering judgment along with user communication based on component criticality is necessary in this situation. Good guidance for tolerancing can be found in the ASME engineering drawing standard Yl4.5M—1982, *Dimensioning and Tolerancing*. (See App. D.) (This topic is discussed in detail in Chap. 5, section on completion of preliminary technical data package.)

Comparison to Available Data

After all measurements have been taken, they must be compared to the dimensions shown in the available technical data. All differences between the stated parameters and the actual dimensions must be noted and recorded. Later in stage 1, after other additional data collection has taken place, these differences are evaluated for significance.

If there is no technical data, as in data development project types, there can be no comparison to available data. Referring back to Chap. 3 (section on determination of samples necessary) for a determination of the number of samples for data development projects, the study of multiple sample parts is needed for the most accurate results to be achieved through reverse engineering. The lack of technical data from which to make comparisons must also be recorded as noteworthy. In the case of full data development the only comparison which can be made is to list the measurement discrepancies between sample parts. These deviations have significance when summarized for the QER.

Disassembly and Assembly Procedures

As the component is being measured it is often being disassembled, although additional sample parts should be reserved for a failure analysis. While disassembling a component, the engineer should prepare a list characterizing each piece and the order in which the pieces are being disassembled. During the disassembly additional design requirements such as torque values or spring compressions will need to be measured. While disassembling a component the engineer should always be on guard for future producibility or reproducibility improvements.

Many singular components consist of inseparable assemblies and subassemblies. These inseparable assemblies are components joined by welding, riveting, epoxy, or any other relatively permanent bonding method. Often they are not meant to be disassembled without destroying or impairing the function of the original component design, and care should be taken in the disassembly of these items. Nondestructive disassembly is useful in determining baseline dimensional characteristics.

There are many reasons to have multiple sample parts. Newer samples are best used as a benchmark versus older samples. A functioning sample part should be kept intact for operational testing. (See section on operational testing, below.) A poor-quality part might be best used for disassembly. Another good reason to have multiple sample parts available for the reverse engineering process is that mistakes can be made, even in the most careful disassembly and material analyses procedures. For the sake of this example let us consider the possibility of differences between a failed part from the working system and a new part from available supply. The older failed part may be bonded in one fashion, and a later model just recently pulled out of stock may utilize a newer bonding method. These differences are noteworthy for the final determination of the best bonding method of this component/assembly, and will impact the disassembly procedures.

In most cases, disassembly procedures are not available and must be constructed from scratch. At best, a proprietary assembly procedure may be available in the factory vault of the original manufacturer. Without this information great care must be paid while disassembling the sample parts. The reason for this is twofold. The most obvious reason is the need for a method to reassemble the component. The reverse disassembly procedure can be reconstituted to build an assembly procedure. The less obvious reason is to develop technical information that heretofore did not exist—for posterity, as it were.

The disassembly/assembly procedure can also provide the reverse engineer some insight into the rationale for the original design. This is no small amount of information if a future attempt at value engineering is to be made. Knowledge of the hows and whys of a part's design allows the engineer to make quantum leaps in thought when seriously contemplating value engineering. If a sheet of metal has hand-drilled holes, it may be because stamped sheet was not available. (Many true-life designs, particularly mechanical designs, are this old.) If the design flow criteria of the original equipment are not jeopardized, the substitution of stamped sheet in the new component may be a viable functional replacement, resulting in the quick resolution of a potentially expensive reverse engineered design.

Material Analysis

With the completion of the disassembly and measurement of all dimensions, the next step is the material analysis and identification. Material must be analyzed before it can be identified. With the emergence of many of today's composites there can be many variations between the assumed material and the actual material. The act of identifying the exact chemical composition of these newer materials has become a specialty in reverse engineering. Material identification could be as simple as a metallurgical analysis of a standard steel or as detailed as the identification of a polymer coating or superalloy. Any variations between the material specified and the material identified should be noted. In some cases the material selected for the original design does not perform adequately for the component application (e.g., the material selected cannot withstand the high temperatures the unit experiences in operation).

Material samples can be difficult to obtain for analysis if the sample part cannot be destructively tested. Choosing the location where sufficient material can be taken for complete and accurate material identification of the part requires an understanding of the function of the part. This is to avoid structural damage to the part in the process of sampling. Expertise in materials science requires knowledge of metals, alloys, surface finishes, coatings, polymers, ceramics, or composites of any multitude of materials. Material analysis and identification are also important in the construction of a bill of materials, whether the part is mechanical or electrical. In electronic and electrically engineered components many parts used in the original assembly are no longer available and the process of identifying a suitable alternative may pose problems when compiling a list of materials. Substitution of parts and materials is addressed in detail in Chap. 5.

Material composition is one of the limiting factors considered in the technical complexity evaluation during candidate prescreening. If a reverse engineering team knows that it does not have ceramic identification skills and cannot access or contract out to obtain them, then the chances for success in reverse engineering are low and the technical complexity is high, resulting in a critical overriding factor to success. This is an example of a case wherein the team capabilities tailor the types of projects that the reverse engineering program can accommodate.

Operational Testing

If operating samples are available that involve moving components and unused samples from supply are also available, then operational testing should be conducted for comparison to any known operating parameters. When there are no moving parts, or the project type is simple product (design) verification, operational testing either may not, or cannot, be conducted. Operational testing is conducted to establish the baseline *operating* parameters for a component. In the case of a valve, testing may be done to determine whether a valve is built for high temperatures or pressures. Cyclic speeds, frequencies, transmission rates, or any other pertinent operational characteristics of the component should be either determined or verified against any design specifications which delineate these conditions. The materials found during material identification should meet or exceed the operating parameters. Failure to either meet or exceed any known operating conditions could warrant a redesign, particularly if safety conditions are not met or are outdated.

Safety factors are not always a matter of judgment. They can often be a matter of legal codes and standards. If during the reverse engineering process as a whole, and stage 1 in particular, a safety or code violation has been found, there may be legal implications. In most cases the reverse engineering focus should then be shifted to value engineering and meeting today's safety criteria.

Discrepancy Review Versus Available Data

The available data is reviewed for completeness while the data adequacy is detailed. The goal is to assure that any data developed in stages 2 through 4 target these inadequacies and discrepancies. Inadequacies are missing data while discrepancies are conflicts between similar data points on a component. The available data forwarded from the prescreen to stage 1 are expected to be complete, although there will be cases where a search for specific additional information is warranted, such as the case of an overlooked test specification.

Available Data Adequacy

The technical data forwarded from the prescreen seldom include all the information needed to remanufacture this part. Most cases of reverse engineering are some form of data enhancement and are therefore missing some data. Recall from Chap. 3 that a product verification project type essentially means that almost all the technical data is complete. If there is some doubt in the mind of the prescreen team that the part can be made with the data on hand, then some form of verification is needed before the part can be remanufactured. In the case of data development, literally no technical data are available. These cases are both extremes, with the vast majority of projects falling into the category of data enhancement. Thus, the available data must be reviewed for both adequacy and completeness. Any missing technical data should have been at least briefly discerned from the prescreen and must be further augmented here in stage 1.

It is easy to state that all detailed drawings need to be developed during reverse engineering when only an assembly drawing is provided from the prescreen. It becomes increasingly difficult to pinpoint what data is missing when a wealth of data is provided from the prescreen. Still, an attempt to identify what information is lacking should be made at this point in stage 1. When this is completed, the data missing is known and can then be targeted for development in stage 2.

Major Discrepancies

In the section, Comparison to Available Data (above), available data were compared to the actual sample part(s) during the dimensional inspection. All discrepancies to the actual component dimensions were listed. These discrepancies fall into many categories. Some are variations between sample parts. Some are variations from the sample parts to the drawing dimensions. Some are material variations or substitutions. In other cases there are wide differences between what is shown on an engineering drawing and what is found on the actual part. Quantifying the importance of these discrepancies is necessary to understand potential design flaws or simple manufacturing differences. The reasons for any differences are many. Determining whether the difference is major or minor is another judgment call. Significant discrepancies should be evaluated in the failure analysis (see next section), which follows next in stage 1.

These differences can be attributed to any number of factors. The original component may have been poorly machined, which may be the cause of high failure rates. The equipment used to produce the original part may be outdated and incapable of producing high-quality parts. Design changes may have occurred since the latest revision date on the available technical data. These design changes may have been instituted without the customer or end user's knowledge. Poor documentation quality control or simply a drafting error may be the reason. (These and other possibilities provide a good argument for including concurrent engineering practices in reverse engineering.) One can never truly know the root cause for design discrepancies, but probable causes may point toward a direction for improving this design. Discussions with the end user or the original manufacturer (when possible) can also provide insight into any data discrepancies.

Significant differences will be noted on the QER, and a summary of major discrepancies will be noted on the stage 1 report. These reports will be discussed in detail later in this chapter.

Failure Analysis

Not all reverse engineering projects will require a failure analysis. If failed parts are forwarded from the prescreen, then it may be necessary to conduct a failure analysis. If few sample parts are available, then conducting an analysis to discern the major failure mode may be difficult. Consistent intolerable component failure under normal operating conditions is a good reason to conduct failure analysis. The failure analysis generally requires study of numerous samples for correct identification of the major failure mode(s). Once the failure mode(s) is (are) identified, it (they) may become the basis for recommending design changes. Design alterations may or may not be tolerated by the user or lead organization and may require approval. If approved, this analysis may be used as the basis for value engineering in stage 2.

Example of Severe Component Failure

There is the case of the binocular pedestal which was introduced to the reverse engineering program because of its consistent intolerable failure. Figure 4-2 shows the original binocular pedestal consisting of cylindrical tubing and welded on fin stabilizers.

The original design was to meet a vibration specification of 50 Hz (cycles per second). The original tubing consistently failed around 38 Hz. The original manufacturer knew of the design's inability to meet specifications and in an attempt to improve the strength of the tubing, inserted a shrink-fit tube inside the inner diameter of the original tubing. This improved the stiffness of the tube somewhat; failure now occurred at 44 Hz, but this was not good enough for field conditions. And so it came to the reverse engineering program, flawed to perfection.

With the failure data in hand, and photographs of other failed binocular pedestals, it was determined that the original tube could not withstand the vibrations because the need to stabilize the structure required that fins be

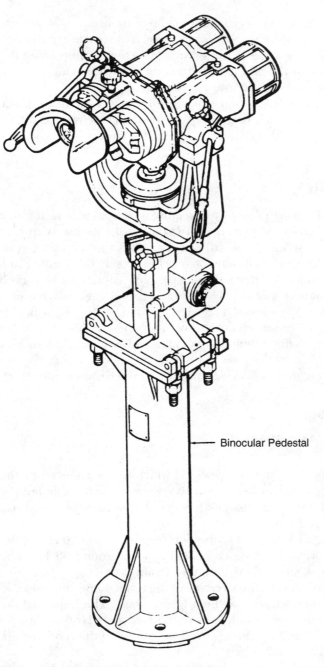

Figure 4-2. Original binocular pedestal design.

welded on, which caused heat stress in the area surrounding the weld and served to weaken the immediate area near the weld rather than strengthen the entire part.

The entire cylindrical tubing was replaced by 1-in-thick square tubing which provided the structural support and eliminated the need for the fin stabilizers altogether. The square tubing could be welded on to an identical baseplate without disturbing the structure of the entire unit. The square tubing then passed the 50-Hz vibration testing with flying colors (actually exceeding 70 Hz with no failure). Figure 4-3 shows the new reverse engineered design with the square tubing on the test stand.

The original binocular pedestal had a unit cost of $910, while the reverse engineered design cost $500, representing a 45 percent unit-cost savings. The entire project, including failure analysis and prototype testing, was estimated at $42,000 and actually cost $42,328. The life-cycle savings (LCS) was estimated to be $126,280. Although the return on investment (ROI) was only 2:1, the design upgrade eliminated the chronic system failures experienced in the field in the past.

Performance criteria and testing data from failed parts should be re-

Figure 4-3. Reverse engineered binocular pedestal.

viewed and discussed prior to any analysis. Care must be taken when evaluating any failure data supplied. Consistent intolerable component failure is an obvious source of failure statistics. However, failure data or statistics are not always a true reflection of failed components. This data may include replacement of annual usage data. Often included in *failure statistics* are components used during routine maintenance procedures, which are thus more correctly characterized as *replacement parts*. (This is the reasoning behind the use of the phrase *replacement rate* in lieu of *failure rate* on the prescreen sheet.) An example illustrating the value of this differentiation might be parts which are scrapped during periodic overhauls of rotating machinery or turbines where replacement of certain critical parts occur during every overhaul. If 15 overhauls are scheduled in one year, then it makes sense to replace specific belts which tend to wear and can cause enormous amounts of maintenance inconvenience if they are not replaced until they fail. This periodic replacement of this type of component has a tendency to artificially inflate annual usage and true failure data.

Failure Analysis in Stage 1 Versus Failure Analysis in Stage 3

Another failure analysis takes place in stage 3 (design verification). Failure analyses in stages 1 and 3 are conducted in the same manner, but that in stage 3 is conducted only on prototypes built in accordance with the technical data generated in stage 2 when these prototypes fail to meet quality and performance criteria. The failure analysis in stage 1 is conducted on failed sample parts forwarded from prescreening.

Up to the conclusion of the failure analysis, the reverse engineering team meeting, visual and dimensional inspections, discrepancy review, and other steps have been a continuation of the data collection begun in the prescreen. The prescreen data collection was conducted on paper data, while the stage 1 data collection is conducted on the physical part in conjunction with the paper data. With the completion of the failure analysis, if required, the data collection phase of stage 1 is complete. Data evaluation on both the physical and paper data now begins.

Quality Evaluation Report Generation

The generation of a QER is a valuable product of stage 1. It is a quantitative and subjective evaluation of the assembly, machining, casting, materials, or other functional characteristics of the component. Its purpose is to make a quality assessment of both the part and the technical data available. It is a review of the overall quality of a component after it has been disassembled, as viewed by the team. There arc instances where the original part is literally a piece of junk. Often poor workmanship is evident in all phases of manufacture and assembly. If, on the other hand, the original part is of high quality, it is less likely that reverse engineering would yield a higher-quality part. There are also cases where a part that appeared to be overpriced in the prescreen evaluation may indeed be a part worthy of its value. If a high-quality part is a candidate for reverse engineering, economics should be an overriding factor. It is important to note here that there are many contradictory, but true-to-life, elements in reverse engineering. None are quite so interesting as the overpriced, low-quality part.

How does one determine whether a component is a quality part? To determine "quality" in this case, the following example questions may need to be answered:

- Does this part exhibit good construction?
- Docs this part match the available data? How closely?
- Are any inappropriate parts evident?
- Are the tolerances befitting the function of the part within the system which it operates?
- Does it do its job well?
- Would I purchase this part for this application if the decision were mine to make?

Let us now review in detail the elements of the QER. Additional elements can be added or tailored to suit company- or department-specific needs from those which appear on this example QER. Figure 4-4 is an example of a typical QER. The project tracking reference number assigned in the prescreen should appear in the upper right-hand corner along with the current date. This information is followed by the item identification data.

Accurate documentation of the reverse engineering project candidate begins with item identification. This is simply the manufacturer's assigned part number (P/N), the manufacturer's preferred nomenclature (item name) for the part, and the name of the actual manufacturer, not the supplier. Often it would be preferable to call a part by its original name, but it is recommended that the manufacturer's nomenclature be used for the item name on all reports. Take, for example, a lag bolt the original manufacturer refused to call anything but a "special screw," which, of course, makes it easier to charge 10 times more for the "special screw" than the lag bolt.

Section II of the QER (see Fig. 4-4) is a quantification of the data availability and discrepancies reviewed earlier. All the information collected

REVERSE ENGINEERING PROGRAM   RPT#_____
QUALITY EVALUATION REPORT   DATE_____

I. ITEM IDENTIFICATION
P/N:_____
Item Name:_____   Mfgr:_____

II. DATA AVAILABILITY/DISCREPANCY REVIEW
        Component_____   Assembly_____
List All Available Drawings   Latest Rev.   Rev. On Hand

Technical Manual Data?

Performance Specifications?

List Of Data Inadequacies:
Mechanical   Electrical
Dimensions: Schematics:
Tolerances: Component Specs:
Finishes: Stable Artwork:
Materials: Operating Parameters:
Interfaces: Test Procedures:
Other: Other:

SUMMARY OF DISCREPANCIES:

MAJOR DISCREPANCIES:

III. FAILURE ANALYSIS SUMMARY (IF CONDUCTED):

IV. PROJECT TYPE DETERMINATION:
Product Verification_____   Data Enhancement_____   Data Development_____

V. OVERALL QUALITY EVALUATION:

VI. RECOMMENDATIONS:

ENGINEERING STAFF SIGNATURES:

Figure 4-4. Sample quality evaluation report.

should be expressed coherently to complete the information requested on the QER. The listing of data inadequacies shown on this example QER is for mechanical parts and should be modified as needed for electronic parts.

Section III requests a summary of the failure analysis if one was conducted. If a failure analysis was not conducted, a line explaining the reasoning for not conducting this is helpful. Section IV requests the identification of the project type. Note whether the project type determined in stage 1 has changed since the prescreen.

The overall quality evaluation of section V should recapitulate all that the reverse engineering team has learned to date about the flaws in this otherwise perfect component. This is a subjective assessment, and a high-quality part in construction, manufacture, and function should be praised as such. Likewise, noting its fatal flaws is a responsibility of the stage 1 reviewing team members.

The recommendations requested in section VI can be as simple as "continue" if all is going as it was anticipated from the prescreen recommendations or as complex as requesting that this part be value-engineered to upgrade its design faults. Remember that a final decision to forward this project to stage 2 is not made on the basis of the QER recommendations but on the full economic, logistics, and quality review with overriding factors shown on the stage 1 report. The final determination to continue is usually made by others with financial responsibilities to the company or department in which you are employed.

Stage 1 Report Generation

Hand in hand with the generation of a QER is the generation of a stage 1 report, which includes key economic and logistics factors not considered while the data is being evaluated in a stand-alone mode. Some of the information needed for the report can be completed while others are reviewing and evaluating technical data. This report must not be concluded until all other steps in stage 1 are completed.

This stage 1 report is the only document that the program decision makers will probably use to make a go/no-go decision. The purpose of this separate report is to summarize all that is known about this singular item to date. An example of an older-style stage 1 report is shown in Fig. 4-5. This document, too, can be tailored to suit program-specific needs by adding or deleting report elements.

Figure 4-6 shows a sample completed stage 1 report from a program in existence in 1988. Note how it differs from the example in Fig. 4-5. This is due to the tailoring described above.

```
┌─────────────────────────────────────────────────────────────────────────────┐
│              REVERSE ENGINEERING PROGRAM                        RPT #_____   │
│                    STAGE 1 REPORT                               DATE_____    │
├───────────────────────────────────────────────────────────────────────────────┤
```

ITEM IDENTIFICATION
P/N:_____
Item Name:_____   Mfgr_____

II. DATA AVAILABILITY/DISCREPANCY REVIEW: (Summary from QER)
      Component:_____   Assembly:_____
Critical Technical Data: (Includes Drawings and Revision Level, Technical Manual and Performance Specifications)

Summary of Discrepancies:

Project Type: Product Verification:_____ Data Enhancement:_____ Data Development:_____

Overall Quality Evaluation: (from QER)

III. PRODUCTION COST ESTIMATES:

Current Unit Cost: $_____   Target Cost : $_____
                                        (Unit Cost X 0.75)
Production Lot Costs:
                                Lot Size   Unit Cost
Small Lot   _____   _____
Annual Requirement   _____   _____
Economic Lot   _____   _____

IV. REVERSE ENGINEERING PROJECT COST ESTIMATES:

            Stage 1 $_____
            Stages 2-4 $_____
            TOTAL $_____

V. RISK ASSESSMENT

VI. OVERALL RECOMMENDATION:

ENGINEERING STAFF SIGNATURES:

STAGE 2 APPROVAL:
Signature:_____   Date:_____

Figure 4-5. Sample stage 1 report.

```
            REVERSE ENGINEERING PROGRAM          RPT #  RE-A66
            STAGE I REPORT
                                                 DATE:  12-07-88
```

1. NSN   2.   NOMENCLATURE
    1H 4370-00-766-8010   SHOULDERED SHAFT

3. PROJECT COST ESTIMATE $ ___3250_____   STAGE I
    (REVERSE ENGINEERING)
                                  $ ___5000_____   STAGE II

                                  $ ___8250_____   TOTAL

4. SUMMARY OF DISCREPANCIES
    Four of the fifty-eight features checked are found to be out of tolerance.
    None are considered to be significant.

5. IS TECHNICAL DATA AVAILABLE? Yes –
    (Pump Drawing No. R-522, Rev. A.)

6. IS TECHNICAL DATA ADEQUATE? No

7. TECHNICAL DATA DEFICIENCIES – The Revision A is not the latest drawing
    revision.

8. QUALITY EVALUATION
    The workmanship and attributes of this part are: Good.
    Dimensional – Four dimensions are out of tolerance.
    Material – Material Analysis reveals part to be made of same material as
    drawing specifies.

9. PRODUCTION UNIT COST   REV ENG   CURRENT
                                           ESTIMATED   UNIT COST ($)
                                          UNIT COST ($)   (NAVSEALOGCEN)
    LOT SIZE

    ONE (SMALL LOT) $ ___1289.00___ $ xxxxxxxxxxxxx

    ANNUAL REQUIREMENT   (85 EA) $ ___427.00____ $ _454.00_____
    (An assumed quantity)
    ECONOMICAL QUANTITY   (350 EA) $ ___388.00____   xxxxxxxxxxxxx

10. RECOMMENDATIONS – The current revision of the drawing is D. The Government
    possesses Rev. A. The OEM has revealed that changes have been made, but
    will not release a copy of the drawings. Recommend proceeding through Stage
    II R.E. because competitive procurement appears economical and a data
    package would permit back-up procurement.

11.
    DESIGN ENGR   PROJECT ENGR   PRINCIPAL ENGR   MECH ENGRG DEPT

    Stage 2 Approved 12/20/88
```

**Figure 4-6.** Sample completed stage 1 report.
```

Let us review the contents of this report, beginning again with the use of the project tracking number assigned in the prescreen and the current date. Section I has a repetition of the item identification data used on the QER. Section II is a summary of the data availability and discrepancy review from the QER highlighting only that information which could influence a final go/no-go decision. Sections III, IV, and V warrant special attention.

Production Cost Estimates

Current Purchase Price and Target Cost.   Noting the current purchase price (from the prescreen unless a significant price change has occurred) on the stage 1 report is critical because this will be the base price for all cost comparisons in evaluating future savings. A target unit cost of 25 percent less than the original cost is calculated by multiplying the unit cost by 0.75. The ability to meet this target cost at the completion of stage 4 is one of the objectives and measures of success in reverse engineering.

Production Lot Sizes.   Production lot sizes and their relative costs must be estimated next. Lot sizing is estimated for three quantities: a small lot of perhaps 1 to 10, an annual lot quantity equal to the annual usage from the prescreen recommendation sheet, and an economic lot. The economic lot quantity is the number of units that would produce the lowest unit price without manufacturing so many that the cost of stockpiling these would exceed any savings gained from such mass production. It is possible that the economic lot size can be smaller than the annual requirement depending on the number of applications of this part in the system within which it operates. The key issues here are what is the smallest production lot size to fabricate and what is a good average lot size to produce. The cost to fabricate in larger lot sizes tends to result in lower unit cost, which is a major goal of many reverse engineering projects.

The relative unit costs associated with these three lot sizes will give the engineer an idea about the value of mass producing this part for system usage. Many procurement algorithms used by individual companies and the federal government recommend or require the purchase of lots which are smaller than is economical and that are meant only to fulfill an immediate need. This information about economic lot purchases can help alter these algorithms to allow the part to be purchased as frequently as possible but in as large a quantity to keep a ready supply available.

The Should-Cost Estimate.   The unit costs associated with the three lot sizes estimated to be the best for this component represent "should costs," otherwise known as the price this item should be purchased for if it were purchased in these lot sizes. This should-cost estimate includes profit margins added to the combined value of the individual components and raw materials plus the cost of assembly. In industry, this "should-cost" logic poses no real problem for items manufactured on a regular basis.

Most people in an industry can readily assess the value of the parts they use in their systems. Most people in procurement functions can understand how the should-cost price becomes inflated due to ineffective purchase requisitions methods. The actual values placed on the stage 1 report are meant to be estimates, but they are also based on a fair working knowledge of the component and component costs for similar products available through other suppliers or manufacturers. There is no real mystery to the should-cost figures once the details of the design are known. It is when we are purchasing black boxes or item numbers of figures on a printout that we become uneducated buyers of the products we use and consume. When we do not own responsibility to purchase parts sensibly we find it hard to care how much we spend doing our jobs. This has caused the plethora of audits in all facets of purchasing which has only served to bring about more rules and regulations that do not allow many to do their job of ordering what is needed in an economic fashion.

This point I would like to illustrate clearly. Most of us will not buy 10 lb of butter at the delicatessen or supermarket merely because butter is on sale. In many households 10 lb would go rancid before it could be used, thereby wasting our personal resources and the value of any real savings per pound. Nor would we buy a car when we must have a truck or van. Our money has a personal value to us, and we are not about to waste it when such large sums are at risk. We buy to suit our needs with a little room to feel extravagant if we can afford it.

In companies and governments there are so many rules governing how we order and purchase items that we cannot use good sense often. If I need 100 nails for a project, I order them from the purchasing department because I know that supplies such as these nails are covered in my project or department funds. If no one else orders nails when I do, then purchasing orders 100 nails from the local nail supply store. Unfortunately, the local supply store carries these nails only in boxes, or lots, of 1000. Purchasing cannot authorize the purchase of the full 1000 nails because that is not what I ordered and my department will not pick up the tab for the additional 900 nails. The supplier cannot find another immediate buyer for the 900 extra nails and refuses to stockpile them until another "strange" purchase request comes through. The procurement person then tries to find another source who can supply in lots of 100. A supplier is found; however, the cost of the 100-unit lot is the same as the cost of the 1000-unit lot. And so a deal is struck between the second source and the procurement office. All contractual details are recorded to prove that no underhanded practices took place on this buy but ultimately, from a macroscopic point of view, the worst deal was made and no one wins. This scenario may be all too familiar to many of you, who may be furious because you have been found out, are laughing hysterically, or may be naive enough to believe that this does not really happen. What we will not do with our own money we gladly do with our company or department funds to stay within the rules that defeat good judgment.

We all tire of fighting windmills, and after a time we give in to all the rules because it is simply too hard to run the gamut anew each day. Here is where reverse engineering is fun. In reverse engineering we have the opportunity to make it right, if only just once—the first time the new part is produced. In reverse engineering ideal parts can be made that stay within the constraints of the system and still operate equally or more economically. Reverse engineering is about getting it right, because what you, the skilled and trained professional, know about your work counts and what judgments you make can affect the final outcome.

Reverse Engineering Project Cost Estimates

The costs associated with the prescreening are considered to be sunk costs, irretrievable, except over the long run when the reverse engineering program is self-sustaining. If, in the prescreen, there appears to be some economic advantage to reverse engineering a component, it will be forwarded for a more thorough review in stage 1. If a project cost tracking system is in place (as recommended in Chap. 3, section on project tracking), then the costs associated with the completion of stage 1 can be quantified, or closely estimated, for inclusion in this report. It will typically cost $1000 or $2000 to a few ten thousands of dollars (or the foreign equivalent depending on labor costs) to complete the steps in stage 1. This cost depends on data availability and the technical complexity of the part as shown in Fig. 3-4.

The cost to complete stages 2 through 4 of the project is an estimate based on all that is known from stage 1 data collection and evaluation. It will include any prototype fabrication and all testing costs. It can be woefully wrong or well underestimated, but the project cost tracking system should be able to sound a warning when the costs are getting out of hand without a reasonable justification such as a change in the scope of the task, the number of prototypes needed to prove the reverse engineered design, or the decision to value engineer.

The costs of stages 1 through 4 define a boundary figure which the project should approach at completion. Experience has shown that final costs vary from project to project but that over time these variations should average out to a median which approaches the original estimates. If they do not, then the procedure used to estimate should be revisited to find the source of the consistent errors in estimation.

The costs to complete stages 2, 3, and 4 can be estimated using a break-down structure based on the following task costs:

- Configuration determination
- Development of drawings
- Development of parts list
- Material determination
- Cost to fabricate prototype(s)
- Establishing testing and/or performance specifications
- Possible destruction of sample part(s)
- Management and logistics support for reverse engineering

At the startup of a reverse engineering program it will initially be difficult to estimate these costs accurately. These costs are closely associated with the complexity levels of the parts or assemblies. On the basis of one organization's needs, the following cost breakdowns were estimated for each task. The relative complexity levels are rated from 1 (the simplest) to 5 (the most complex). If a level 1 part is a hose or circuit card with three resistors and a level 5 part is a pump or a magnetic amplifier, the following may be used as a guideline until in-house costs can be more accurately determined:

Configuration determination = level 1—$150/part
(level 3—$2000/part)
(level 5—$8,000–$10,000/unit)

Development of drawings = $700/drawing (including revisions)

Development of parts list = $150/item

Material determination = $250/part

Cost to develop prototypes = $ should-cost estimate

Establish test and performance specs = level 1—$500/part
(Level 3—$3000/part)
(Level 5—$10,000/unit)

Destruction of sample parts = $XXX unit cost × units destroyed

Management-logistics support = 15% overall cost

Summing up these costs may give a fair estimate of the cost to complete stages 2 through 4. With experience these estimates are expected to be refined. The total cost of stages 1 through 4 must ultimately be amortized over the remaining life cycle of the system equipment for this project to be economically feasible. The final return on investment uses the actual project costs in calculations, and so the estimates should be as accurate as possible. As the reverse engineering program matures, the accuracy of the estimates will improve.

These estimates figure strongly in the final decision to continue or stop work at the conclusion of all stage 1 evaluations. There are two reasons for this. If, when all stage 1 work is completed, it is the best estimate of the lead engineer that this item cannot be manufactured for 25 percent less than the current unit cost *and* economics was the overriding factor for introducing this item as a reverse engineering candidate, then there is little hope to reverse engineer a more economical part or recoup the costs incurred to date. If economics is not the overriding factor but obsolescence is, then savings calculated on a unit-cost basis cannot be performed since it is too difficult to place value on an irreplaceable part.

Risk Assessment

This is simply a general engineering judgment of whether there is enough potential for a return on the investment in either dollars, marks, or yen, or in system maintenance needs met, that would otherwise go unmet. As a general rule of thumb high-speed rotating aircraft parts would constitute a high risk whereas a simple stationary system component operating at a low temperature might be a low-risk reverse engineering project.

Overall Recommendation

This is the reverse engineering team's collective chance to make a case for forwarding the project to stage 2 or discontinuing it with the completion of stage 1. First, it must be determined whether the available technical data is adequate and if they are to be used in their present state to fabricate future spare parts. The recommendation is to compete future acquisitions to the technical data reviewed in stage 1.

Second, it must be determined whether it is economical to proceed to stage 2. In this case a brief review concerning the adequacy of the available data is needed, something along the lines of "Revision (rev.) A is available, rev. C is the most current. Changes from rev. A to rev. C are unknown." This should be followed with a one-sentence statement about the economics such as "Unit costs are anticipated to decrease 50 percent after reverse engineering, project costs are reasonable, and projected ROI could be as high as 35:1." The collective recommendation is to continue reverse engineering in stage 2 or to terminate the project with little hope of return. If the potential for ROI is low or questionable, there is the option to give this project a low priority, especially if better candidates are being reviewed, and revisit this project after the better candidates have been exhausted.

The recommendations will usually fall into the categories of compete, low priority, terminate, or proceed to stage 2. There is one additional option, although it rarely occurs. This happens when the candidate points to grave system problems. This could indicate that the presumed life cycle of the entire system is much shorter than management expects. This shortened life cycle may indicate that the time to junk the entire system is near and reverse engineering may not be recommended. In this case the recommendation would be to seriously examine total system replacement or modernization. Besides, who better than those most familiar with the operating system could make this determination with hard data to support this claim? If management truly adheres to employee empowerment, then this recommendation will be taken seriously. It is entirely possible that management has a separate agenda and this recommendation cannot be accommodated at this point in time but may be kept on a back burner until a better opportunity exists for the company to take action.

If the team's recommendation is not accepted, there may be a chance to renegotiate the project. If this is not an option, then the decision by the approval body must be accepted and other projects begun.

The Go/No-Go Decision

The stage 1 report contains the data necessary for the project leader to recommend a decision to go on to stage 2 or to justify the decision to stop the project at the conclusion of stage 1 on the basis of information collected and reviewed to this point. After stage 1 evaluations are completed, a fully informed engineering decision concerning the cost and risk involved in stages 2 through 4 can be made. Either it is, or is not, worth the potential cost, in terms of supply support issues or from an economic standpoint. The costs incurred in stage 1 are considered sunk costs and are nonrecoverable. This is the price tag to find out all anyone should need to know in order to make as fully informed a decision as possible before proceeding further into stages 2 through 4, or continuing to purchase this item from the same or another source of supply.

The final decision always rests with the organization which is funding the reverse engineering program, not the project leader, however a joint decision of agreement is recommended.

It is of note that at any point in the prescreening, or during any of the four stages a reverse engineering project could, and sometimes should, be stopped. At any time after the completion of stage 1 an unknown or un-knowable overriding factor could be discovered. Prior to stage 1 the real costs (i.e., the costs for prescreening, data collection, and preliminary evaluation) are minimal. Stage 1 costs are not recoverable unless the pro-ject continues all the way through stage 4 and a ROI is realized. It is least expensive to discontinue a flawed project before starting stage 2.

There is the case of the military vehicle component which was in stage 3 when the lead engineer, who had not followed this project closely, realized that there was not enough data to build a prototype. The project was al-ready over budget, and the additional cost to finish the project would in-crease the total cost 50 percent. It was also learned at that point that the part cost after reverse engineering would double instead of decrease. Had the available information been evaluated well early in the process and the warning signs been flagged, there would have been no need to decide be-tween asking for additional funds to finish the project and deciding to stop it far into stage 3, as it would have been known that there would be no ROI.

Needless to say, these are professional and moral decisions no engineer would like to face. This example serves to reiterate that thorough and accu-rate groundwork should be conducted. Despite the up-front price tag, this work outweighs the long-term consequences of mistakes further into the project.

Others who have conducted reverse engineering programs have com-plained that the sunk cost of thorough prescreening is wasted or is too high a price to pay for the selection of high-quality reverse engineering candi-dates. All too often these same programs which skimped on prescreening lost money in the long run, and most are defunct. It is wise to think about the price of the investment and the cost of doing it wrong before wasting resources in stages 1 through 4.

5

Stage 2:
Technical Data
Generation

Technical data generation is every design engineer's dream come true. All the preliminary work has been completed and the remaining work has been budgeted and approved. The feeling is not unlike that of the house painter who has spent weeks preparing a house for the new coat of paint by scraping the old paint off the surface. The painter has painstakingly scraped and sanded all the moldings and window sills, covered all the surfaces which will not be painted, and assembled and prepared the tools of the trade. The painter does not wait until the moment the brush is in hand to check to see if it has been properly cleaned before applying the new coat of paint. The tools are on hand, clean, and ready for use. In essence, all the preparation has been done. The budget to complete this project has been agreed on and is available. The painter's assistants have reviewed their role in the project. Work can now begin.

The objective of the work at hand is to develop a complete and unrestricted technical data package which will be sufficient for both the fabrication and procurement of the part in the future. The engineering drawings will define the configuration of the component, and the technical data package will contain all additional inspection and quality assurance requirements to manufacture a like part.

The technical data develops throughout many iterative phases in the four stages of the reverse engineering process. At this point (the beginning of stage 2) in the four-stage process there is the data accumulated from the prescreen along with the data collected and evaluated in stage 1 to make the preliminary decision of whether to proceed with reverse engineering. The flow diagram in Fig. 5-1 shows the progression of data development in sequential form. In stage 2 the missing or inadequate technical data will be generated as engineering drawings and incorporated in a *preliminary drawing set* (PDS). Once the PDS drawing set includes the associated performance specifications, quality assurance requirements, and testing criteria, it becomes the stage 2 *preliminary data package* (PDP). After passing the requisite in-house staff reviews this information becomes the *preliminary technical*

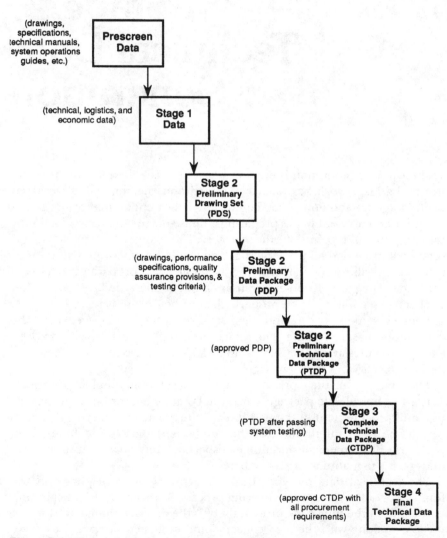

Figure 5-1. Technical data development flow diagram.

data package (PTDP) ready for transmittal to stage 3. When prototypes are built and pass both the operational and system testing in stage 3, the PTDP will become the *complete technical data package* (CTDP) at the conclusion of stage 3. In stage 4 the procurement requirements are added to the CTDP to become the *final technical data package*. Stage 4 is not concluded until the engineering and economic report and prototypes are delivered with the final technical data package for signature approval.

With this overview of data development, the serious task of developing the missing technical data is begun in stage 2. Figure 5-2 provides an overview of the steps involved in technical data generation.

In this chapter some frequently used phraseology regarding drawings and technical data is reviewed. Terms such as *engineering drawing* and *technical data* may seem trite to the experienced user of these types of information, but the international world of engineering is rapidly requiring the redefinition of what is considered to be standard practice. By revisiting what we *think* we know, we can either verify that we know everything or relearn something we previously overlooked, and it is easy to overlook the simplest of things in our daily lives. For effective communication in this new order there must be some common ground for mutual understanding so that all parties can apply the same meaning to these common terms.

Consider for a moment the computerized world we live in where the boundaries between hardware and software are rapidly disappearing. In the not-so-recent past, the interchangeability of phraseology caused much confusion, and today standard methods for understanding, transmitting, and communicating data are being developed to fill the need for compatibility

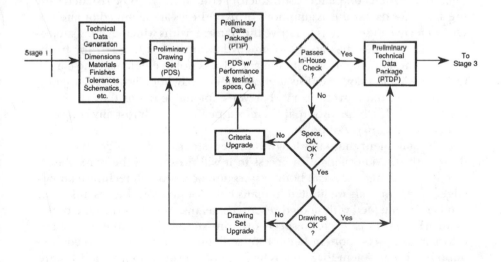

Figure 5-2. Overview of stage 2 (technical data generation).

among all computer vendors' wares. It is equally important that we engineers in the land of "real hardware" all understand each other; therefore, while some concepts may seem basic, we must start with the same definitions to achieve results which will mean the same thing to all who rely on our work.

The Importance of Developing Unrestricted Technical Data

It is important that the company or department which conducts reverse engineering strive for unrestricted data as an end product. One hallmark of the future will be our collective ability to communicate and share information. If a component has been selected as a reverse engineering candidate because of a lack of supply support for an operating system in one company, there might be other existing systems which could share from the benefit of the technical data generation. The primary desire of any corporation is to recoup the costs of reverse engineering. Fighting the urge to hoard the technical data developed is critical. Corporate hoarding of technical data has generated the need for reverse engineering in the first place. If the technical data were already available, even in some obscure place, reverse engineering would not be necessary.

Let us digress just a moment to discuss the potential for sharing information as a strategic corporate goal. In today's environment there is more information on any given subject than any single human being needs. We limited humans can barely process the available data we have and make sense of it. Often completely contradictory conclusions can be reached with the same raw data if the assumptions or hypotheses are changed or altered. One objective in science is to derive the natural truths which overlay subjective reality. Science has access to much data in its noble quest for truth. Scientists and engineers are often made of less noble elements. The corporations which employ scientists and engineers may not always be searching for any truth, but a profit from whatever exploitable truths they can manage to extol. Truth be known, if there is a profit in it, almost any company will pursue the power of profit.

If the statement that knowledge is power is considered to be true, then those with the "knowledge" or access to it will prosper in the years ahead. Scientists and engineers can both possess and access much technical knowledge. In recent years we limited humans have found that which is unknown can best be pursued with pooled resources, teams, or consortia. One of the hallmarks and most critical aspects of team success is the ability to share information. A component being reverse engineered is a problem component. If it is a problem for me, it is likely to be a problem for you. If I share what I know with you, then maybe you will share something with me, and in this manner we build a teaming relationship which we both hope will profit us in the future. If I share with you, and you exploit me, I do not suppose I will team with you or assist you again and I will gain all the future profit from my work. Only if I share my knowledge can you profit also.

Pooling resources and sharing information is one of the keys to success in these days of tight budgets and limited resources. The R&D (research and development) budgets must be spent wisely, and reverse engineering is not meant to cut into these budgets but to be self-sustaining and provide a return on investment in some reasonable period of time. The need for R&D investment is so great in so many technical areas, especially manufacturing. There is so much more to learn that multiple organizations do not need to be doing R&D, reverse engineering, or other types of technical development on the same items at the same time. If, while searching for candidates, you discover that four other companies use the same unsupported component, is it not possible, even wise, to propose that you will develop unrestricted technical data for a system component that all four companies are having difficulties obtaining for a percentage of the cost to reverse engineer? Or better yet, you could trade your reverse engineering service for some of their technical resources, whether personnel or equipment—high-tech bartering—anything but information hoarding. This way no one company profits in the short run and everyone profits in the long run. Let us leave our children a legacy of information sharing and international cooperation. (*Note:* Developing unrestricted technical data does not mean developing unrestricted, nonproprietary engineering drawings and slipping in restricted testing requirements. This violates the principles of teaming and information sharing and the essence of cooperation.)

This radical concept of technical information sharing rubs against the corporate grain of profitability. While one may not want to apply this concept to original designs in order to avoid compromise of trade secrets, design patents, and the like, it should be considered for reverse engineered parts. This is because these represent problem components, not production parts manufactured on a regular basis. It is not yet profitable to manufacture reverse engineered components on a regular basis; only to extend this support to operating slowly deteriorating systems. The need for interchangeability and compatibility of products may point to information sharing as a corporate strategy in the future of the manufacturing world as in the world of computers, but it is recognized that this day is not yet here.

To reinforce this concept, keep in mind that once upon a time analog and digital systems were completely separated by the inherent design restrictions imposed by the selection of either one or the other base operating mode. Today even our telephones, the most basic and common analog system, is a digitally operating piece of equipment with associated digital transmission systems. This transformation from analog to digital has been completed with simple conversion algorithms. Information sharing via telephone lines has brought us the facsimile (fax) machine and immense changes in our use of telephone lines to transmit data from our computers. These radical shifts in our ability to share information may soon bring similar changes to the world of manufacturing and preparing for it with information sharing on non-production-oriented parts may be a way to introduce the future to today's corporate goals and strategies.

Technical Data Generation

The technical data which needs to be generated in stage 2 includes the dimensions, the materials, the surfaces, the finishes, the interfaces, the tolerances, the performance and testing specifications, and the quality assurance requirements which will verify that the reverse engineered part will at least fulfill the function of the original component—if not outperform its predecessor. The technical data should include any special tooling and its associated documentation. The explicit location of reference points for machining purposes must also be stated.

A generic *technical data package* (TDP) provides a technical description of an item adequate for engineering and logistics support, an acquisition strategy, and producibility. The description defines the required design configuration and procedures required to ensure adequacy of item performance. It consists of all engineering drawings and associated parts lists along with all performance criteria, test specifications, and applicable standards. This will also include the life-cycle fatigue testing, nondestructive test and evaluation criteria, first article testing and inspection criteria, and quality assurance provisions and packaging details, particularly if the item has a limited shelf life.

The task of developing a complete data package is no small feat. The emphasis in the previous sentence is the word *complete*. To develop only the engineering drawings is to have generated the preliminary drawing set (refer back to Fig. 5-1). Only with the addition of the performance specifications, test, inspection, and quality assurance information verified in stage 3 will the TDP be complete. A thorough review of the requirements of a complete military- or defense-related TDP can be found in the military specification MIL-T-31000, *General Specification for Technical Data Packages*. Although this specification was developed for military applications, its principles as applied by commercial organizations can constitute an effective method to ensure completeness in all engineering drawings.

Those unfamiliar with military or government specifications may find it useful to obtain copies of this and other government specifications or industry standards referenced in this text to familiarize themselves with the language and interpretations of these and like documents. Much of the work engineers do must conform to either international, industry, or government/military standards. This trend for conformity in engineering drawing practices, applications, and interpretations is on the rise, although there is a growing practice of adopting the nongovernment standards whenever practicable, or combining government and industry standards when this results in a superior set of technical requirements.

The use of standards is increasing worldwide, as is the need to effectively communicate and interpret these documents. The standards and specification cited in the Appendixes are fairly common and global in application. In keeping with the many efforts worldwide to promote and encourage the use of the standard principles involved, it is recommended that these and other primary referenced specifications and standards be reviewed as an effective tool for engineering design, production, and quality control which can provide both technical and economic advantages for all parties. These documents apply to data prepared by either manual or automated methods, such as computer-aided design–computer-aided manufacturing (CAD/CAM) systems, or combinations thereof. They also apply whether using U.S. customary units of measurement, the International System (SI) of units, or combinations thereof.

The TDP consists of many elements. The military specification MIL-T-31000 is for the general specification for TDPs and prescribes the requirements for preparing a TDP composed of one or more TDP elements and related TDP management data products. The following is a listing of TDP elements.

1. Conceptual design drawings and associated lists
2. Developmental design drawings and associated lists
3. Product drawings and associated lists
4. Commercial drawings and associated lists
5. Special inspection equipment (SIE) drawings and associated lists
6. SIE operating instructions
7. SIE descriptive documentation
8. SIE calibration procedures
9. Special tooling drawings
10. Specifications
11. Preservation, packaging, packing, and marking data
12. Quality engineering planning list

13. Software and software documentation

14. Test requirements documents

Those items referring to TDP management data products are

1. Source control drawing approval request

2. Drawing number assignment report

3. Proposed critical manufacturing process description

4. TDP quality control program plan

5. TDP validation report

The TDP elements conforming to the requirements of the specification MIL-T-31000 are intended for use as a basis for design evaluation, competitive acquisition, installation, maintenance, modification, or engineering support. The TDP management data products are intended for use by the acquiring activity to ensure that the TDP elements will be satisfactory for their intended use.

Developing Engineering Drawings

Perhaps the most important form of technical data in the entire data package is the set of engineering drawings and associated parts lists developed to physically describe the component and establish the configuration baseline. Engineering drawings are also commonly known as *build to print* or *level 3* drawings. The reverse engineering team has conducted the dimensional inspection flawlessly and compared the actual dimensions of all available operational and nonoperational samples before considering a physical dimension accurate enough to commit to the engineering drawing. Whether the engineering drawing is a hand-drawn sketch, a blueprint, or a CAD/CAM program, every physical dimension and detail must be precisely and accurately described. The validation of dimensional accuracy is the reasoning behind the frequent measurement of multiple sample parts. Guidance for the development of complete and accurate engineering drawings that will be uniformly interpreted can be found in the ASME standards Y14.24—Types and Applications of Engineering Drawings; Y14.34—Parts Lists, Data Lists, and Index Lists; Y14.26—Digital Representation for Communication of Product Definition Data; and Y14.5—Dimensioning and Tolerancing, excerpts of which can be found in Appendixes A, B, C, and D respectively.

Dimensional Accuracy

True dimensional accuracy is difficult to achieve. An example bushing may be useful to illustrate this point. (Refer to Fig. 5-3.)

The external diameter dimension for the front view of the bushing will be a single number measured out to a certain number of degrees of accuracy. For the purposes of this example, accuracy within $\frac{1}{1000}$th will be used. The external dimension represents the outer diameter of the entire length of the bushing. Over the length of the bushing the diameter measurements may vary from an average of 2.025 (inches or centimeters) to 2.005 or 2.050. The bulk of 100 sample measurements along the entire 8-in length of the bushing lie within the range of 2.013 to 2.047. The large number of samples taken over the 8-in length would allow the average engineer to assume that stating the dimension to be 2.025 ± 0.020 would be valid. If by chance only two sample measurements were taken to validate the dimension and these happened to be 2.035 and 2.045, then the average engineer would probably state the diameter dimension to be 2.040 ± 0.005.

With only two sample measurements this would appear to be true. The trouble would lie in the reality of manufacturing a part to the 2.040 measurement in the diameter over a part which is truly 2.025 units in diameter. The difference of 0.015 unit over the entire length of the part would be disastrous if the bushing had to be inserted into a component that had an internal dimension of 2.035 ± 0.005. Not one single part manufactured to this dimension of 2.040 would fit into the higher assembly.

Validating dimensional accuracy is a point well taken if one uses this example as a learning tool. Dimensional accuracy is a relative measure in truth. The "real" dimension varies from one set of points on the component to another set of points. The average of the entire, albeit infinite, set of points along the length of the part, which in this case is the bushing, would be a fair representation of the true dimension. In reality, although many measurements may be taken, the highest and lowest variations are considered to be the outer limits with the mean measurement a reasonable

Figure 5-3. Bushing example for dimensional accuracy.

representation of the true dimension. This is true with the exception of very poorly machined parts.

Although it is extremely helpful to have multiple sample parts with which to compare dimensional data, it is more advantageous to have a slightly outdated, nonproprietary drawing revision to use as both a reference and a reality check. A nonproprietary higher-level assembly drawing showing the interference dimensions for a component would also be valuable.

Any competent engineer could potentially go insane in the quest for precise and accurate measurements of all dimensions. For this reason most reverse engineered drawings require that a shop fabricate the component to the reverse engineering drawing generated in order to confirm that the part can actually be manufactured to these dimensions and that the component can operate in the higher assembly without causing systemic failure.

Engineering Drawing Types and Applications

To quote ASME Y14.24M—1989, *Types and Applications of Engineering Drawings,* an *engineering drawing* is an engineering document that discloses (directly or by reference) by pictorial or textual presentations, or combinations of both, the physical and functional end-product requirements of an item. An *item* is a general term used to denote any unit of product or data, including materials, parts, subassemblies, equipment accessories, computer software, or documents which have entity. A *part* is an item made from a single piece of raw material or from multiple pieces joined together which are not normally subject to disassembly without destruction or impairment of the designed use (e.g., transistor, screw, gear, wheel bearing). *Component* and *part* are used synonymously in this text. A *parts list* is a tabulation of all parts and bulk materials (except those materials that support a process) used in the item. Referenced documents may also be tabulated on a parts list. Items listed on a subordinate assembly parts list or specified in a referenced document are not generally repeated on this first-tier listing. A parts list can contain information on alternate sources for hard-to-find parts. A parts list is also known as a *bill of materials,* a *list of materials,* a *stock list,* or an *item list.* An *end item,* or end product, is an item, such as an individual part or assembly, in its final or completed state.

Drawing definitions are intended to permit preparation by either manual, computer-aided, or photographic methods. Using the common definitions just delineated, we are now prepared to define the accepted standard drawing types used to establish engineering requirements. ASME Y14.24M—1989 and MIL-STD-100E dated 30 September 1991 (which supersedes DoD-STD-100) are the best first-tier reference documents available, with the International Standards Organization (ISO) TC10/SC1 (Technical Committee 1, Subcommittee 5) establishing corresponding guidance for the international community. The following types of engineering drawings are the most frequently used to establish engineering requirements.

1. Layout drawings
2. Detail drawings
3. Assembly drawings
4. Installation drawings
5. Modifying drawings
6. Arrangement drawings
7. Control drawings
8. Mechanical schematic drawings
9. Electrical and electronic diagrams
10. Special-application drawings
    a. Wiring harness drawings
    b. Cable assembly drawings
    c. Printed-board drawing sets
    d. Microcircuit drawings
    e. Undimensioned drawings
    f. Kit drawings
    g. Tube bend drawings
    h. Matched set drawings
    i. Contour definition drawings
    j. Computer program and software drawings

It is important to pay attention to which types of drawing establish item identification and which do not. *Item identification* is the part number, identifying number, or descriptive identifier for a specific item. In the government arena the manufacturer's Commercial and Government Entity Code (CAGEC or CAGE Code) is also needed on drawings for complete item identification.

layout drawing: Depicts design development requirements. It is similar to a detail, assembly, or installation drawing, except that it presents pictorial, notational, or dimensional data to the extent necessary to convey the design solution used in preparing other engineering drawings. A layout drawing generally does not establish item identification.

detail drawing: Provides the complete end-product definition of the part(s) depicted on the drawing; establishes item identification for each part depicted thereon.
    monodetail drawings: Delineate a single part.
    multidetail drawings: Delineate two or more uniquely identified parts in separate views or in separate sets of views on the same drawing.

assembly drawing: Defines the configuration and contents of the assembly or assemblies depicted thereon; establishes item identification for each assembly. Where an assembly drawing contains detailed requirements for one or more parts used in the assembly, it is a detail assembly drawing.

installation drawing: Provides information for properly positioning and installing items relative to their supporting structure and adjacent items, as applicable. This information may include dimensional data, hardware descriptions, and general configuration information for the installation site.

modifying drawings: Include altered-item, modification, and selected-item drawings. These are not used for items made from raw or bulk materials, items purchased in bulk lengths (extrusions, channel nuts, hinges, etc.), or such semiprocessed items as blank panels, castings, or electronic equipment drawers, which use detail or detail assembly drawings.

    altered-item drawing: Delineates the physical alteration of an existing item under the control of another design activity or defined by a nationally recognized standard. This drawing type permits the required alteration to be performed by any competent manufacturer including the original manufacturer, the altering design activity, or a third party. It establishes a new item identification for the altered item.

    selected-item drawing: Defines refined acceptance criteria for an existing item under the control of another design activity or defined by a nationally recognized standard which requires further selection, restriction, or testing for such characteristics as fit, tolerance, material (in cases where alternate materials are used in the existing item), performance, and reliability within the originally prescribed limits. This drawing type generally permits selection to be performed by any competent inspection or test facility including those of the original manufacturer, the selected design activity, or a third party. Also establishes a new item identification for the selected item. Although visible physical modification is not performed, the item is, because of the selection technique employed, demonstrably different from other items which meet only the requirement imposed on the original item.

    modification drawing: Delineates changes to items after they have been delivered. When required for control purposes, a modification drawing shall require reidentification of the modified item.

arrangement drawing: Depicts the physical relationship of significant items using appropriate projections or perspective views. Reference dimensions may be included. An arrangement drawing does not establish item identification.

control drawings: There are five types of acceptable control drawings.

    procurement control drawing: Provides criteria for performance, acceptance, and identification of supplier items by disclosing the engineering design characteristics required for (1) control of interfaces and (2) to ensure repeatability of performance.

vendor item drawing: Provides engineering description and acceptance criteria for purchased items, a list of suggested suppliers, the supplier's item identification, and sufficient engineering definition for acceptance of interchangeable items within specified limits. The vendor item drawing number appears with any applicable dash number(s) for identifying item(s) on engineering documentation; the administrative control number may be marked parenthetically on the item, in addition to the supplier's item identification.

source control drawing: Provides an engineering description and acceptance criteria for purchased items that require design-activity-imposed qualification testing and exclusively provides performance, installation, and interchangeability characteristics specifically required for the critical applications. It includes a list of approved suppliers, the supplier's item identification, and acceptance of items which are interchangeable in the specified applications. Also establishes item identification for the controlled item(s). The approved items and sources listed on a source control drawing are the only acceptable items and sources.

design control drawing: Discloses the basic technical information and performance requirements necessary for a subcontractor to complete the detailed design required to develop and produce an item. The design control drawing specifies the unique identifier of the item; the item identification may be that assigned by the subcontractor's design disclosure drawing (if known) or an administrative control number assigned by the design control drawing.

interface control drawing: Depicts physical and functional interfaces of related or cofunctioning items. It does not establish item identification.

identification cross-reference drawing: An administrative-type drawing which assigns unique identifiers that are compatible with automated data-processing (ADP) systems and item identification specifications; provides a cross reference to the original incompatible identifier.

mechanical schematic diagram: Depicts mechanical and other functional operation, structural loading, fluid circuitry, or other functions using appropriate standard symbols and connecting lines. This is a design information drawing and does not establish item identification for the item(s) delineated thereon.

electrical and electronic diagrams: In accordance with ANSI Y14.15—1966, *Electrical and Electronics Diagrams* or ANSI/IEEE (American National Standards Institute, Institute of Electrical and Electronic Engineers) STD 991, *Preparation of Logic Circuit Diagrams,* depict the elements or functions of electrical or electronic items using standard symbols. These diagrams do not depict items to scale. They are design information drawings and seldom establish item identification for the item(s) depicted thereon. There are seven types of electrical and electronic diagrams.

functional block diagram: Depicts the functions of the major elements of a circuit, assembly, system, etc. in simplified form.

single-line diagram: In accordance with ANSI Y14.15, depicts the course of an electrical or electronic circuit, or system of circuits, and the elements thereof using single lines, symbols, and notes. A single-line diagram conveys basic information about the operation of the circuit, but omits much of the detailed information usually shown on schematic diagrams.

schematic diagram or circuit diagram: Depicts electrical connections and functions of a specific circuit arrangement without regard to the physical shape, size, or location of the elements.

connection diagram or wiring diagram: Depicts the general physical arrangement of electrical connections and wires between circuit elements in an installation or assembly. It shows internal connections, but may include external connections which have one termination inside and one outside the assembly. It contains the details necessary to make or trace connections involved.

interconnection diagram: Depicts only external connections between assemblies, units, or higher-level items.

wiring list: Consists of tabular data and instructions necessary to establish wiring connections. A wiring list is a form of connection or interconnection diagram. When the wiring list includes materials and such material is not called out on the assembly drawing, the wiring list establishes the item identification for the wires as a bundle or kit of wires.

logic circuit diagram: Depicts the logic functions of a system at any level of assembly.

special-application drawings: There are 10 types; details of each of these can be found in App. A.

Examples of each of these types of drawings can be found in ASME Y14.24M—1989. Appendix A lists each type of drawing along with their individual application guidelines and requirements. It is important to be familiar with this and with the associated drawing specifications pertinent to your specific equipment.

The corresponding U.S. government standard for engineering drawings was previously identified as the military standard MIL-STD-100E. An accurate perception of DoD engineering drawing practices necessitates user recognition of ASME Y14.24M—1989 and ASME Y14.34M—1989 for *Parts Lists, Data Lists and Index Lists,* as being a composite set. MIL-STD-100E includes chapters entitled "Preparation of Engineering Drawings" (Chap. 100), "Types of Engineering Drawings" (Chap. 200), "Drawing Titles" (Chap. 300), "Numbering, Coding and Identification" (Chap. 400), "Markings on Engineering Drawings" (Chap. 500), "Revisions of Engineering Drawings" (Chap. 600), and "Associated Lists" (Chap. 700).

Tolerance Determination

Dimensional tolerancing should conform to the nationally recognized standard in ASME Y14.5M—1982, *Dimensioning and Tolerancing.* (See App. D.) This standard covers dimensioning, tolerancing, and related practices for use on engineering drawings and in related documents, and establishes uniform practices for stating and interpreting these requirements. Internationally, this standard is similar to ISO/TC10/SC5, *Dimensioning and Tolerancing* (International Standards Organization, Technical Committee 10, Subcommittee 5), although differences do exist.

Geometric dimensioning and tolerancing is a means of specifying design and drawing requirements with respect to the actual "function" and "relationship" of part features. This technique, when properly applied, ensures the most economical and effective production of these features during fabrication. Thus geometric dimensioning and tolerancing can be considered both an engineering design drawing language and a functional production and inspection technique. The use of this standard promotes uniform understanding and interpretation by design, production, and inspection personnel. Uniform practices within these functional areas allows for consistent and competitive practices when applied to component production.

This standard is important primarily because it is economical and facilitates the production of interchangeable, complete, uniform components that mate (interface) with their associated parts on a consistent basis.

The following definitions of *dimensioning* and *tolerancing* will be needed.

dimension: A numeric value expressed in appropriate units of measure and indicated on a drawing and in other documents along with lines, symbols, and notes to define the size or geometric characteristic, or both, or a part or part feature.

basic dimension: A numeric value used to describe the theoretically exact size, profile, orientation, or location of a feature or datum target. It is the basis from which permissible variations are established by tolerances on other dimensions, in notes, or in feature control frames.

reference dimension: A dimension, usually without tolerance, used for information purposes only. It is considered to be auxiliary information and does not govern production or inspection operations. A reference dimension is a repeat of a dimension or is derived from other values shown on the drawing, or on related drawings.

tolerance: The total amount by which a specified dimension is permitted to vary. The tolerance is the difference between the maximum and minimum limits.

unilateral tolerance: Tolerance in which variation is permitted in one direction from the specified dimension.

bilateral tolerance: Tolerance in which variation is permitted in both directions from the specified dimension.

geometric tolerance: The general term applied to the category of tolerances used to control form, profile, orientation, location, and runout.

A review of the fundamental rules clearly defines the engineering intent of dimensioning and tolerancing and shall conform to the following:

1. Each dimension shall have a tolerance, except for those dimensions specifically identified as reference, maximum, minimum, or stock (commercial stock size). The tolerance may be applied directly to the dimension.

2. Dimensions for size, form, location of features shall be complete to the extent that there is full understanding of the characteristics of each feature.

3. Each necessary dimension of an end product shall be shown. No more dimensions than those necessary for complete definition shall be given. The use of reference dimensions on a drawing should be minimized.

4. Dimensions shall be selected and arranged to suit the function and mating relationship of a part and shall not be subject to more than one interpretation.

5. The drawing should define a part without specifying manufacturing methods. Thus, only the diameter of a hole is given without indicating whether it is to be drilled, reamed, punched, or made by any other operation. However, in those instances where manufacturing, processing, quality assurance, or environmental information is essential to the definition of engineering requirements, it shall be specified on the drawing or in a document referenced on the drawing.

6. It is permissible to identify as nonmandatory (optional) certain processing dimensions that provide for finish allowance, shrink allowance, and other requirements, provided the final dimensions are given on

the drawing. Nonmandatory processing dimensions shall be identified by an appropriate note.

7. Dimensions should be arranged to provide required information for optimum readability. Dimensions should be shown in true profile views and refer to visible outlines.

8. Wires, cables, sheets, rods, and other materials manufactured to gage or code numbers shall be specified by linear dimensions indicating the diameter or thickness. Gage or code numbers may be shown in parentheses following the dimension.

9. A 90° angle is implied where center lines and lines depicting features are shown on a drawing at right angles and no angle is specified.

10. A 90° basic angle applies where center lines of features in a pattern or surfaces shown at right angles on the drawing are located or defined by basic dimensions and no angle is specified.

11. Unless otherwise specified, all dimensions are applicable at 20°C (68°F). Compensation may be made for measurements made at other temperatures.

The primary building blocks for geometric dimensioning and tolerancing are based on the geometric characteristics and symbols listed in Fig. 5-4 on the next page.

Appendix D gives additional guidance in the use and application of this standard.

The Substitution Factor

Most components proposed for reverse engineering are manufactured to designs at least 5 to 10 years old, as a minimum, with many as old as 30 or 40 years, maybe more. The rapid rate of technological progress has left many types of standard design techniques in the dust, so to speak. To repeat a design technique which today's engineer has learned from and yet improved on would be both unprofessional and irresponsible in some cases. In these special cases the positive net return on investment from a design substitution would be advisable, provided it is done after notifying either the person requesting or the person authorizing the reverse engineering. Any changes to a design will require some level of approval unless no other alternative is available. Once the approval has been obtained, this part will require that a prototype be produced for performance and system testing to validate the proposed design alteration. Until an upgraded design with substitutions has been tested successfully and the new design drawings have been signed off, this upgrade is considered to be only a proposed solution to the technical problem posed for reverse engineering.

American National Standard
dimensioning and tolerancing

ANSI Y14.5M-1982

	Type of tolerance	Characteristic	Symbol
For individual features	Form	Straightness	—
		Flatness	⏥
		Circularity (roundness)	○
		Cylindricity	⌭
For individual or related features	Profile	Profile of a line	⌒
		Profile of a surface	⌓
For related features	Orientation	Angularity	∠
		Perpendicularity	⊥
		Parallelism	//
	Location	Position	⌖
		Concentricity	◎
	Runout	Circular runout	↗ *
		Total runout	↗↗ *
* Arrowhead(s) may be filled in.			

Figure 5-4. Geometric dimensioning characteristics.

This discussion of the issues of substitution includes components, materials, or even manufacturing techniques. Without substitution there is the possibility of duplicating original design flaws. There is the mechanical example of the case of a rotating part that continually failed after approximately 7500 h of use. The part was brought into a shop for reverse engineering. An exact duplicate in dimensions, materials, and heat treatments was produced and tested. During performance testing the reverse engineered part failed near the 7500-h mark, also.

Casing Example. A casing problem for a variable resistor involving an expensive welded casing for some electrical components comes to mind. The casing had a wall mounting bracket as the backplate. The engineer handling the unit wondered if the back plate was really necessary considering the expense in reproducing the casing. Rather than blindly pursue duplication, the question of the actual usage of the variable resistor was posed. After a few days of searching for a knowledgeable user it turned out that the unit had never been mounted to a wall, but was used as a handheld unit in the field. With this new knowledge the engineer placed the components in a $10 plastic case in lieu of the $100 bracketed case, duplicating the function while never adversely affecting the final product or system. Figure 5-5 shows the bracketed casing for the variable resistor on the left, while the handheld reverse engineered unit is on the right.

The original unit cost was $141, while the reverse engineered unit cost was reduced to $60, representing a unit savings of 57 percent. The project cost was estimated at $6500, but the actual cost increased as a result of this research and was $8998. The life-cycle savings was calculated to be $88,218 with a return on investment of 8.8:1.

This example serves as a reminder that reverse engineering is not "xeroxing" a design. In most cases xeroxing a product design is valid reverse engineering; in other special cases product design improvement by substitution is valid. When using substitute components, one does not disturb the functionality. A very small percentage of reverse engineering projects will be-

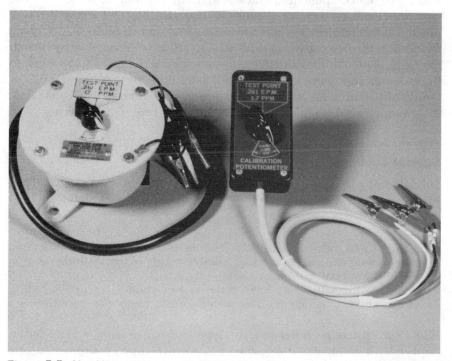

Figure 5-5. Variable resistor casing substitution example.

come significantly improved on, and thus become value engineering projects.

Electrical and Electronic Components.   The special case of component substitution is directed primarily toward electrical and electronic components. Without some amount of component substitution, virtually no electrical or electronic component can be reverse engineered. As stated earlier, many designs are quite old and design techniques can be outmoded. In no other area of technological advances have parts so quickly become obsolete as in electronics, with the possible exception of computers or cold fusion. In electronics it is not a case of design fads but improved performance capabilities of the products themselves or the manufacturing processes that produce these parts.

Most electrical or electronic designs will require some level of product substitution. In some cases it will be quite extensive and will upgrade the design so completely that the part can be called a *value engineered part.* Many more cases will simply require that one type of resistor or capacitor be replaced with a newer, more capable resistor or capacitor. In many of these cases there are companies which publish catalogs with component equivalencies, so the component upgrade is a given, a known. In some cases the part which will be substituted must be searched out by looking through much product information and comparing the original performance specifications to the newer part specifications to ensure that an adequate match has been made. To find an adequate substitute is the usual procedure; to find a better alternative is designing for the future, not repeating the past. One of the aims of an engineer is to design what is necessary. To create a design which achieves this and more is the art of reverse engineering.

Special Tooling

Special tooling is occasionally used in reverse engineering projects. MIL-T-31000, the military specification for technical data packages, defines special tooling as

> unique tooling which is mandatory to manufacture an item. It differs from tooling designed to increase manufacturing efficiency in that the use of the special tool imparts some characteristic to the item that is necessary for satisfactory performance and cannot be duplicated through other generally available manufacturing methods.

If any special tooling is needed, the type and name of the special tool is required. An example of a special tool may be a computer chip design generation kit.

Completion of a Preliminary Technical Data Package

Now that the proper type of engineering drawings have been developed as a preliminary drawing set (PDS) and the associated performance, testing, and quality assurance requirements have been added, along with any additional special requirements such as special tooling, the preliminary data package (PDP) is readied for verification. Referring to Fig. 5-2, on overview of stage 2, the next step is to conduct an in-house review of the PDP for accuracy and completeness. This is necessary before this PDP can be forwarded to stage 3 for prototype fabrication.

This PDP review should be conducted with an eye toward manufacturability to improve the chances for first time success in prototype fabrication. Although the fabrication shop will be sure to come knocking at your door if there is a problem with the design furnished from stage 2, it is the responsibility of the design engineers to provide the best information possible to those who must construct a part to the reverse engineered design. The primary responsibility of the shop personnel is not to bring basic design errors to your attention, but to suggest ways to improve the design. Information such as the location of the primary points of reference for machining are critical because they can expedite efficient manufacture of a part. Improperly selected machining references can both cause errors in all dimensions which depend on these reference points and slow the entire fabrication point down with redundancies and inefficiencies.

If the PDP does not pass an in-house check, a review of the performance, testing, and quality assurance specifications will be easiest because it will not require that the drawings be revised. A criteria upgrade is usually all that is necessary. If the PDS is inadequate, then a revision is required. After either a criteria upgrade or a drawing revision, another in-house check must be conducted. Any number of iterations of this process can be conducted. Only when both the drawings and criteria are correct can the preliminary data package be considered a preliminary technical data package (PTDP) which is then passed on to stage 3 (design verification).

6

Stage 3: Design Verification

The proof of the pudding is in the eating; the best way to ascertain whether something turned out as originally intended is to sample or test it. The proof in reverse engineering is in the strength (accuracy and validity) of the cumulative data developed in the previous stages. *Proof* establishes validity and the quality or state of having been tested, or proved. The design generated in stage 2 is verified by testing of a preliminary technical data package (PTDP) on one or many levels.

The primary method used to validate the PTDP is to build prototypes in accordance with only the stage 2 engineering drawings and to then test these prototypes against the criteria developed in stage 2. This verification method requires that two separate operations be conducted, yet the outcome is dependent on the completeness of the data to achieve success in both areas. The two operations are prototype fabrication and prototype testing. An overview of stage 3 is shown in Fig. 6-1. This overview has both the fabrication and testing displayed sequentially after the PTDP has been approved and it is determined that prototypes are necessary.

A Preliminary Technical Data Package Approval

Prior to prototype fabrication and testing in stage 3, the PTDP requires approval by a technical entity outside the reverse engineering team. In most

107

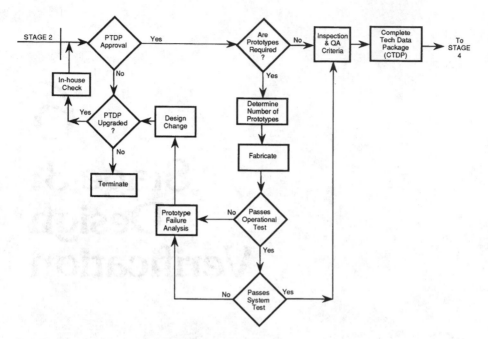

Figure 6-1. Overview of stage 3 (design verification).

cases the reverse engineering design team will not be the final technical authority on reverse engineering projects. After the completion of an in-house reverse engineering team design review on the PTDP of drawings and associated specifications, these should be submitted to the organization which has the technical approval authority. The technical authority in many cases is the end user, system supervisor, or production manager, who will have the authority to grant the technical approval. The reasoning behind this second level of approval is because this person, or these people, will be responsible for the implementation and maintenance of the system in which the reverse engineered component will be installed. They may also be instrumental in ensuring that the reverse engineering information be used for the purchase of future units. If this alternate source of technical approval does not like using the reverse engineering data, chances are that the part will be purchased the "old" way even after stage 4 approval. Approval from the appropriate agent or organization is necessary to provide a macroscopic view of the part within the system.

Approval of the PTDP is the first step in stage 3. Is the PTDP approved? If not, then it must be upgraded by iterations until it can be approved. The approval agent or organization can suggest additional technical, testing, inspection, or quality assurance requirements which may have been inadvertently overlooked by a purely design-oriented group. Any potential adversarial relationship from the past between the technical design and technical approval organizations should be put aside for the common good of the future of the systems which are currently lacking adequate supply parts. In some cases the oversight activity is a technical management organization responsible for both operations and funding. With any luck these responsible people have technical backgrounds, connections with the field users, and friends in procurement. With all these facets they can act to assist in furthering this PTDP in lieu of acting like a bottleneck and endangering the future of the project.

No matter who the approval agent or organization is, it is not to be considered a trifling activity or an unnecessary step in the process of reverse engineering. These same people may become the resident "reality check"; with their innate system knowledge and an understanding of the criticality of this part, they will either approve the PTDP or recommend revisions. Revisions should serve to improve the end product and should be accommodated after full and open discussions which end in agreement by both parties. Review and technical criticism should be viewed as a service. Only after completion of all the necessary upgrades to the PTDP (those which can be known) prior to any prototype testing can approval to continue with stage 3 can be granted. Without this approval, the project hangs in limbo and may be terminated. This could happen in a case where the criticality of a component was grossly underestimated in the previous stages of reverse engineering.

Prototype Determination

The next step in stage 3 is to answer the key question "Are prototypes required?" This is one measure of risk in reverse engineering. If prototypes are *not* built, the risk increases. Building and testing prototypes often constitutes a large portion of the cost to reverse engineer a part. While weighing the cost of building and testing the prototypes against the risk of adverse affects may appear to be a negative way to measure the risk prior to the finalization of any design, the responsible engineer is required to question the design adequacy of any part. Measuring risk is an issue that can be determined only on a case-by-case basis by those most familiar with the system, not in this text for all parts and for all systems.

If prototypes are not required, the PTDP should be rigorously reviewed before approving it as a complete technical data package (CTDP) and concluding stage 3. There are a few isolated reasons for not requiring prototypes to verify a reverse engineering design. A product verification project may have proved that the available technical data were adequate even with a minor alteration or two. The sample parts worked in the system and the low risk assumed in not testing a prototype will not jeopardize the final outcome of the project. If the part is mechanical in nature, many sample parts were used to develop the technical data, and there were considerable available data to compare measurements to, and there are no moving parts; prototype fabrication would be superfluous. The nature of the part can determine whether building a prototype is advantageous.

Closely related questions concerning prototype determination are "Can this design be tested?" (a "yes" or "no" answer should suffice), followed by "How can this design be tested?" (Presumably this was determined in stage 2.) An item with no moving parts may not be operationally testable, but testable only in a working system. Figure 6-2 shows two mechanical parts, a nozzle ring and a bushing, which could be prototyped but not operationally tested. They could have been system tested; however, the risk of not building prototypes was low and in each case no prototypes were built.

A prototype can always be built, but it cannot always be tested. How could a bent tube be tested? A simple circuit design may only require bench testing to known operating parameters to verify its design adequacy. A proto-

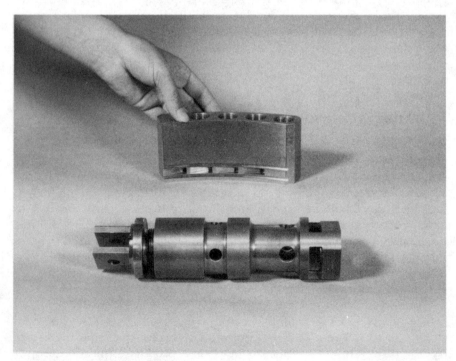

Figure 6-2. Bushing and nozzle ring example parts.

type may be system-testable but not operationally tested. A test fixture may not be available to test a prototype at high temperatures, pressures, or speeds. The cost to build a suitable test stand or fixture may be exorbitant. Outside independent testing laboratories and facilities may be needed in cases where the equipment is unavailable.

If there have been any design changes, alterations, modifications, or deviations from any material benchmarked in either the prescreen or stage 1, then the potential risk of not building and testing a prototype must be weighed. When there have been changes since the prescreen and stage 1, then the part does require testing. Is this a low-risk or high-risk part? A *high-risk part* is one in which failure will potentially cause loss of life, catastrophic loss, major equipment damage, or long-term downtime for the system.

If the answer to any of the questions concerning risk points to design evaluation and verification by means of prototyping, then the next question is "How many?" From a practical and economical standpoint, it is recommended that low-risk mechanical parts have between one and three prototypes built to verify the design which evolved from stage 2 technical data generation. Ordinarily, 5 to 10 electrical or electronic parts should be fabricated. The reason for this increase in prototype fabrication for reverse engineered electronic or electrical parts is that procuring the necessary parts to build prototypes is as economical for 10 electrical or electronic components as for one mechanical part based on material costs. The cost of a single resistor is almost the same as the cost to purchase a minimum or sample lot; resistors are a case in point. Sometimes with electrical parts free samples can be obtained, thereby reducing the cost to build prototypes. Ultimately, these 10 electrical or electronic prototypes can be added back into the supply system as spares if testing goes well, producing a net economic advantage for the system even after the cost of building 10 prototypes is taken into account.

The number of prototypes fabricated in stage 3 should increase with the risk of serious failure. Consider, for example, a high-pressure valve. If a valve has an operational rating of 6000 psi [pounds per square inch (lb/in^2), or metric equivalent], the pressure testing of the prototypes is usually $1\frac{1}{2}$ times the rated pressure for safety reasons. In this case the pressure test would be conducted at 9000 psi. If a critical component of a piece of rotating machinery normally operates at 5000 revolutions per cycle then testing at 7500 revolutions per cycle would be the minimum necessary to ensure the manufacture of a safe component. For safety reasons multiple prototypes should be tested if high temperatures, pressures, or speeds are involved.

This text does not delineate which factors of safety are to be used. Industry standard safety factors apply equally to reverse engineered parts and new designs. This brings up a major point concerning risk in reverse engineering. It is wise to never bend standard rules of design and testing for parts in the reverse engineering process.

As the safety factors and quality assurance requirements (due to technical complexity) increase, so do the number of prototypes which need to be fabricated and tested. There is a fine line between the cost to fabricate multiple technically complex parts and the number needed to ensure the safety of operations which can be determined only by the organization conducting the reverse engineering.

Prototype Testing Requirements

Once the number of prototypes has been determined (typically 1 to 3 for mechanical parts and 1 to 10 for electrical parts), they are to be built in accordance with the drawings and specifications developed in stage 2. An additional question now must be addressed: "What level of testing is needed to assure a quality design?" There are two primary levels of prototype testing. These are *operational* testing and *system* testing. Project type can influence the extent of testing. A data enhancement project usually will require less extensive testing than a data development project. A critical or technically complex assembly will require more extensive testing than almost any other part in reverse engineering.

If sample parts are available, they should be tested along with the prototypes to establish that the prototype performance is equivalent. Documenting the results of the original and prototypes should be included in one report for each type, and level of testing is an invaluable comparison when judging the adequacy of the reverse engineered design.

Operational Verification to Original Parameters

Operational testing verifies that the reverse engineered design conforms to the same design parameters as the original part. What is commonly known as *bench testing* of the operational characteristics of a part is one of the lowest levels of testing. Bench testing is simply a check to verify that the part performs to the minimum operational requirements, i.e., that the input current generates the desired output voltage. Following bench testing are cyclical testing, environmental testing (temperature, humidity, salt spray, burn-in, etc.). There are many varieties of operational testing. The specific performance characteristics of the reverse engineered part to be tested must be determined for each individual project.

Sometimes it is impossible to know what the operational requirements and conditions are for the original component. In cases such as these it is very helpful to have some knowledge about the system in which the part operates. If this information exists, it should have been included in the data collection conducted during prescreening. Without any performance data to compare it to, any operational or bench testing can be conducted only with good technical judgment and comparative sample parts.

System Testing

The real proof of a reverse engineered design lies in the system testing. The part must perform its function as evaluated from a systemic viewpoint. In other words, if it does not work in the system that it is designed for, it does not work. For system testing the reverse engineered component is installed into a working system to validate that it can perform the function of an equivalent part. Arranging for a system test is not the effortless task it appears to be. Often a system test must be planned long in advance. Sometimes there are long lead times for the placement of prototypes in operating systems. It is advantageous if the same person who approves the PTDP also can set up the system testing. This can help minimize the time it takes from PTDP approval to prototype fabrication to system testing.

If an experimental system is available, placing the prototype in this situation lowers or avoids any risk to operating systems in the field. Sometimes there are working testbeds which provide a good proving ground for the design work. Ideally, if multiple prototypes are built, then multiple prototypes should be system tested. A time limit for testing needs to be established, and then the part should be removed from the test system and reevaluated by inspection for any degradation.

Prototype Reevaluation

The operating prototype(s) has (have) passed all operational and system testing. After a specified test period the part should be removed from the system in which it is being tested for reevaluation purposes. Reevaluation of the tested prototype is conducted against the same inspection criteria for the production-level item. Any degradation in the physical condition or operating characteristics of the prototype should be noted for review in the failure analysis.

Prototype Failure Analysis and Redesign

If a part does not pass its operational or system testing, then it is back to the drawing board; that is, it is time to conduct a prototype failure analysis and correct any and all defects before completing the TDP using this reverse engineered design. Sometimes this is an easy task when the problem is apparent. Sometimes it takes a great deal of expertise to pinpoint the mode(s) of failure. In very few cases it becomes obvious that there are hidden, proprietary design features that could not be discerned in the entire reverse engineering process to date. This is where there is a risk that the design upgrade will not be approved and the part will be terminated from the process.

Following any redesign due to operational or system failure, the design change must be resubmitted for approval; if it is not approved either because it has changed the function or is any way unsuitable to the approving organization, then the risk of termination looms. Projects types such as product verification and data enhancement run the lowest risk of failing the testing requirements. Data development project types or projects with a high level of technical complexity, being heavily design-intensive, require the most stringent design and testing.

If the design changes are approved, the prototype determination and testing cycle begins again. If additional prototypes are necessary, it must first be determined whether the original prototypes can be upgraded accordingly or if new designs need to be developed. The cycle continues until an approved prototype can pass both operational and system testing.

Inspection Criteria

Inspection criteria are added to the PTDP at this point in stage 3. The inspection criteria will be used for future production lots of this part. While no errors are tolerated in the prototype fabrication, a limited number of defective parts are allowable in a production mode. Most often a sampling method and procedure is chosen in accordance with corporate or agency procedures. Specific design features will need to be verified as accurate on every unit produced using this technical data. Choosing which key features must conform on every part, and their associated tolerances, must now be added to the PTDP. Once inspection criteria are specified, future production lots must meet the inspection requirements now listed in the PTDP. Inspection reports should confirm that the inspection criteria have been met as well as specified.

Quality Assurance Criteria

Addition of quality assurance criteria must be similar to that of inspection criteria. Quality assurance requirements are also needed for future production lots. Corporate or agency standard quality assurance procedures are adequate to fulfill this requirement. Once a prototype has passed all its testing requirements, the design has been proved. This proven design will later become a production level item similar to all other system components, requiring similar criteria. For internationally marketed components, the ISO 9000 series will act as the overall quality assurance guidelines while in military applications MIL-Q-9858, *Quality Program Requirements* and MIL-I-45208, *Inspection System Requirements* are the standard specifications for engineering equipment. Industrial manufacturers will have similar level specifications also. This information should be added to the PTDP in an appropriate manner. Like the inspection criteria, the quality assurance criteria are both specified and met. This is substantiated in any quality assurance reports delivered with future production lots.

The Complete Technical Data Package (CTDP) Review

If any errors in the engineering drawings or testing procedures surface during the fabrication or testing of the prototype, the shop personnel should have notified the lead engineer as soon as possible to correct the drawings. By the time you reach this last step in stage 3, many of the engineering drawings and testing procedures have received multiple revisions and modifications during the many reviews. The final reverse engineering technical data submitted for the CTDP should all show a revision level of A or 0. This primary revision-level labeling of A or 0 will make it easier to identify this as a reverse engineered item and serves to set it apart from older in-house design data. All in-house iterations should have been tracked through a separate procedure.

With all upgrades embedded into the latest revision of the PTDP, along with the addition of inspection and quality assurance requirements, the PTDP is about to become the CTDP (complete technical data package). This CTDP is the entire design package which will be the basis for all future production of this part. Final preparation for stage 4 means having one last CTDP review by the shops, engineers, and senior-level design team members. Completion of a stage 3 design verification checklist is recommended. A sample stage 3 design verification checklist is shown in Fig. 6-3.

This checklist, like all other reverse engineering documents, can be tailored to suit specific project needs but should include the minimum information pertinent to a drawing review, technical and performance specifications, testing criteria, inspection and quality assurance provisions, and "the final question."

When all this information has been assembled and reviewed, as well as double-checked, the TDP is now complete. Procurement provisions are added in stage 4 since they have little to do with the design verification of stage 3. The entire design package is then considered to be ready for stage 4 (project implementation).

116

Stage 3 Design Verification Checklist

It is important that all the questions listed below be answered affirmatively before proceeding to stage 4—project implementation. Additional project specific information can be included to verify the completeness of the technical data package as necessary.

Drawing review

- Is there a listing of all drawings and technical data?
- Is revision A or 0 the latest version of each drawing?
- Does the information on the parts list or bill of materials specified match the final prototypes?
- Are all dimensions, finishes, and tolerances included?
- Are schematics and artwork an accurate representation of the final component?

Technical and performance specifications

- Are there design or performance specifications which must accompany this technical data package?
- Do these specifications match current industry or government specifications?
- What are the operating parameters for this item?

Testing criteria

- Are both the operational and system testing criteria specified?
- Is the requirement to pass these criteria clearly specified?
- Do final copies of the testing reports and data substantiate that all requirements have been met?

Inspection and quality assurance provisions

- Are specific inspection and quality assurance criteria required?

The final question

- Is the information contained in the complete technical data package sufficient to fabricate, test, inspect, and procure this part from another manufacturer?

Figure 6-3. Stage 3 design verification checklist.

7

Stage 4: Project Implementation

The preliminary technical data package (PTDP) assembled in stage 3 meets all the test, inspection, and quality assurance requirements needed to verify the design. The stage 3 design verification checklist has been completed and all technical data generated is in its final form. With the inclusion of all test, inspection, and quality assurance reports, the data have collectively become the complete technical data package (CTDP) and are forwarded to stage 4 for finalization and project implementation. An overview of stage 4 is shown in Fig. 7-1.

Some important steps in the finalization of the TDP are included in stage 4 and must not be overlooked. To the casual observer or the purely technical individual some of these steps may seem superfluous, but each provides some necessary level of guarantee that this project will be implemented without incident and provide the part specified on the drawings. The addition of procurement requirements assures that all technical requirements will be complied with and all testing will be both completed and documented in future procurements. The delivery of prototypes, the engineering and economic summary, and the CTDP with procurement requirements, form the final TDP. To gain final signature approval, a formal presentation of this cumulative information should be made to those who can grant the final approval for the use of this information as the basis for future procurements. The formal presentation will provide the approval body with the opportunity to ask questions concerning the project and its various nuances as well as provide the engineering lead with the opportunity to showcase the capabilities available in-house.

117

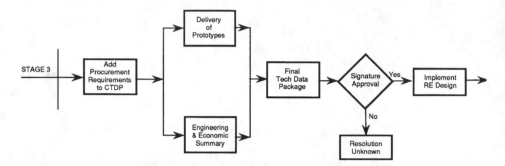

Figure 7-1. Overview of stage 4 project implementation.

The Addition of Procurement Requirements to the CTDP

Procurement requirements are a separate entity from design requirements and as such do not belong in stage 3 (design verification). Some of the procurement requirements will be derived from some of the technical requirements in the CTDP, such as source control for special parts of an assembly. The CTDP with the additional procurement requirements will be used as the basis of all future procurement actions after receiving stage 4 approval as a final TDP. (*Note:* At this point the final TDP may be referred to as the *TDP*; however, *TDP* is technically the acronym for the final product after approval. The interim stages of reverse engineering have been given individual acronyms to differentiate the various stages of the technical data development.)

Every corporate and government procurement activity has its own way of handling the dissemination of technical data to those who would fabricate, manufacture, or supply parts to a company or organization. A large company has one set of rules where a small business might have less restrictions. There will be organizations which produce the parts using facilities available in house; therefore, no additional procurement requirements will be necessary. No matter what fabrication and procurement method is used, it is virtually guaranteed to vary from organization to organization. Even the best-designed reverse engineered part must eventually be purchased to be of value to the organization; therefore, the procurement activity must have a hand in the finalization of the TDP.

Procurement requirements may include restrictions on the sources which may bid on the CTDP. A component which includes a bearing may have restrictions which apply to the selection of bearing manufacturers on a source control drawing included in the CTDP. A bearing manufacturer may be prevented from bidding on a valve assembly. A valve manufacturer may not be allowed to bid on electronic parts. Procurement requirements will almost invariably include some form of procurement restrictions or compliance issues; therefore, the role of procurement becomes important to the finalization of the CTDP.

If a part is used on a system which will have aerospace applications, certain specifications will apply that would not if the part were to be used in conventional manufacturing environments. Likewise, the environmental clauses for a piece of equipment for a naval system will include salt exposure testing and a high-humidity testing requirement that will guarantee the continuing operation of the part in any seawater-exposed environment from the arctic to the equator.

Procurement requirements may include specifications, packaging, packing, shipping, handling, delivery dates, forms of payment, compliance testing, first-article testing, inspection, and reporting procedures. The procurement personnel will guarantee that the TDP has all the necessary requirements to fulfill the mission of the part. If the engineers in stage 2 have fully understood the application of the part, they have made the job of procurement infinitely easier by specifying which requirements have associated inspection or testing and which must be requested in the bid package. The procurement specification will be based on the CTDP but will take on a life of its own in stage 4.

The Delivery of Prototypes

The delivery of prototypes is an important step because it provides the visual proof of design sometimes overlooked in stage 4. At least one functioning prototype should accompany the final TDP during the formal presentation for signature approval. If possible, one or more of the original sample parts should accompany the prototype for comparison. Although all tested functioning prototypes are valuable assets, it is seldom necessary to deliver all prototypes in the signature approval presentation (if multiple prototypes have been made and tested) because it is only redundant and serves little purpose in proving the validity of the reverse engineered design. Any remaining prototypes will need to be inducted into the supply system after signature approval is obtained.

If more than one set of prototypes were fabricated in stage 3 as a result of failure or significant degradation, all sets of prototypes should be forwarded to stage 4 to accompany the CTDP as it is prepared for finalization. These multiple prototypes can be made available as proof to the naked eye of the casual reverse engineering participant who has funded this effort for the past few months or years, that yes, all technical difficulties have been overcome by design improvements, and verified by repeated testing. It is assuring to those who would approve a design package to see the damage to an initial prototype the first set of field trials caused, and the resultant lack of damage to the final prototypes. To see only a final product after much trial and tribulation has been passed through is to minimize the effort required to improve the end product. To the practiced eye, these iterations do not serve to highlight your misjudgments or mistakes, but serve to prove that your persistence has truly solved a difficult design problem. So deliver all prototypes to stage 4, even the mistakes (assuming that it is possible that reverse engineers make more mistakes than the original designers). Of this lot a few example parts should be selected for the formal signature presentation, while the remainder can be easily made available should a need arise. On delivery of prototypes, the sample parts supplied in stage 1 are also included unless they have been destroyed through disassembly or testing.

The prototypes produced in stage 3 that have passed system testing are proven end products and can now be used as valuable assets to the system for which they were designed. The introduction of proven prototypes from stage 3 to the supply system brings a much needed maintenance function benefit in the form of future supply parts. These reverse engineered prototypes can be considered to be an economic savings in and of themselves. This saving, although often small compared to the project cost, can add up to large savings over the course of a large reverse engineering program with many projects.

The following hypothetical example illustrates this point. The International Big Money Corporation of America, Inc., has reverse engineered 100 parts of a large system over a 5-year period.

If 50 of these parts are mechanical and on the average 2 prototypes are fabricated and have passed system testing, there will be 100 supply assets to add back into the corporate supply system. If the unit cost of these parts average $100 each, then the corporation has saved $10,000. There is also the intangible cost savings of not having to complete the paperwork that is associated with any purchase, and there is no waiting period to ordering and receiving these parts in the supply system. That could mean the difference between a short downtime for the repair of a critical machine and a 3-week wait to rush-order the part from the supplier (who often will add a service charge for rushing the order).

Of the 50 electrical and electronic parts reverse engineered, only an average of 5 prototypes are submitted with the final TDP. These 250 parts normally retail for $100 each, also. This is a $25,000 real savings. Taken together, the $35,000 saved can fund additional projects or be reinvested in the reverse engineering program to help it become financially self-supporting. In this way prototype delivery can be of direct benefit to the company in an immediate fashion in terms of supply support and economic return on investment.

There is one final issue in prototype delivery. If the tested prototypes are an asset, does that make the original parts already in the supply system and those on back order a liability? If the original parts are found to be so deficient that they are rendered useless, then you have a potential disposal problem on hand if these parts contain hazardous, toxic, or radioactive materials. In the vast majority of cases there should be no serious problem with the continued use of original components until exhausted. There should be no reason to interrupt the supply system to introduce reverse engineered components, but if there will be a negative effect to scrapping or disposing older parts, be certain that this information is included in the engineering and economic report.

The Engineering and Economic Report

Before the approval body endorses the entire project, procurement regulations, prototypes, and so on, the reverse engineering team has one final task: to generate a final engineering and economic report. The responsibility for the generation of this report rests primarily with the lead engineer. The report functions as the executive summary of the project and provides the documentation and economic justification for reverse engineering the part—from prescreen through stage 4. This document is not just a technical summary but a business document from which good decisions can be made. No example report is provided in this text since this should not be a single page form and should always be unique to the project in all ways, except perhaps in its basic format.

Each engineering and economic report details the individual processes used to achieve the end result of the each reverse engineering project. A standard format for this report includes background information detailing the reasons for selecting this item as a candidate, the establishment of the final configuration of the part, a summary of the technical data development in stages 2 and 3, a summary of major difficulties overcome in design and testing, a summary of all testing, and the economic results of reverse engineering. It should read like an executive summary of the individual reverse engineering project and should "sell" the reverse engineered design or solution to the approving body, especially if the reverse engineered design truly is better than the original design.

This report also functions as a crucial management tool because the effectiveness of each reverse engineered project can be measured in real dollars (or pesos, yen, or marks). The effectiveness of the investment made and the expected return on investment from each project is recorded in these engineering and economic reports. A cumulative record, over time, of individual project successes by the reverse engineering team can validate the benefit of continuing support for the program.

For reference purposes a copy of the original prescreen recommendation sheet and stage 1 report are recommended as attachments. Many of the figures calculated, such as the life-cycle savings, will be recalculated and compared to the pre–stage 1 figures. These will also become important in the determination of the final return on investment. These reports can also be used to roughly measure the time it has taken to complete this project, from start to finish, a pertinent management metric to some organizations.

Background Information

The background information should include the primary reason for selecting this item of supply as a potential candidate during prescreening, such as economic reasons, obsolescence, and lack of supply support, and whether anything changed or augmented this recommendation in stage 1. If this item was suspected as the underlying cause of much system downtime, and this was the major reason for introducing this item, it is also of note. Basically, discuss anything of note which justified the selection of this item as a candidate and its acceptance as a project.

Configuration Control

Configuration control is the means of establishing the final baseline design of this item and its associated part number. Any original manufacturer part numbers will be correlated to any new part numbers established as a result of reverse engineering. A new part number may need to be established to differentiate the reverse engineered item from the original item of supply, although the various numbers should be cross-referenced. Future configuration control and definition will rest with the reverse engineered design after the first procurement of this item to the final reverse engineered TDP. If there are any substantial differences in the original and reverse engineered designs, this should be discussed to the extent necessary to establish the final reverse engineered design as the future design. A part may have multiple system applications. Not all system applications may accommodate the reverse engineered part for a variety of reasons, and this should also be noted. This section of Engineering and Economic Summary may also contain a listing of important industrial or government/military specifications and standards used in the development of the technical data package.

Summary of Technical Data Development

The report should include a brief summary of the reverse engineering design process including the number of sample parts provided with their associated physical condition, the number of prototypes produced, and the design problems encountered and overcome. A brief review of the number and types of engineering drawings developed which are necessary to define the part is helpful. The summary of technical data development is a record of the various tasks needed to complete the project and will give management the insight into the real-life tasks that were performed to bring this project to completion.

Most jobs appear easy to the outsider, especially when they are only looking at the finished product. How many of us think of all the tasks performed to build our cars? We might appreciate the effort that the person who installed our car's brake or electrical system, if we could comprehend all this person has done on our behalf. This report is then a final opportunity to recognize the extraordinary efforts that a particular department or shop contributed to the project.

Summary of Major Difficulties Overcome

If any extraordinary obstacles presented themselves and were overcome, this is important to document in the engineering and economic summary. If, for instance, the material composition derived from the material analysis matched no known alloy, and three other material identification methods were needed to pinpoint the material, this would be considered to be a difficulty overcome. If no testing facilities were easily available and a test stand had to be fabricated to accommodate the testing of the prototype, this should be discussed. If outside assistance was needed for verifying the operating conditions, this, too, should be noteworthy. A particular person may have tracked down a user to understand a seemingly arbitrary design requirement which turned out to be a substantial design detail, and this also is a major difficulty overcome.

Summary of Testing

A listing of all tests performed, from bench testing through extended system testing, should be the basis of the testing summary. Any known conditions should be listed, and any comparative findings from the actual testing should be discussed. If the performance of the prototypes surpassed that of the sample parts, be certain to note this. Deviations from expected results can be included in the summary. This should be a comprehensive indication of the extent of testing, if any, performed to verify that the prototypes work at least as well as any sample parts. Again, if special test equipment was necessary or had to be fabricated, this should be included in the report. The passing of all testing requirements establishes the true validity of the reverse engineered design. The summary of testing should do so, also.

The Economics of Reverse Engineering

The economics involved in the reverse engineering decision should provide the final justification for the decision to fully implement the reverse engineered design. Preliminary economic data should be compared to current economic indications. The only case where economics is overlooked, but still discussed, is obsolescence. For the vast majority of projects the economic portion of the report is the most important part of this document.

Projected versus Actual Costs.   From the stage 1 report the projected reverse engineering costs for the completion of stage 1 and the costs of stages 2, 3, and 4 should be compared to the actual costs at the conclusion of the project. The estimated new cost to manufacture the component is compared with the cost of the original component both in real dollars (pesos, yen, or marks) and as a percentage of the original cost. Should-cost and actual amounts will be compared if possible.

If the original part cost was $537 and the reverse engineered part should cost $372, the $165 difference will have more meaning expressed as a 30.7 percent decrease in unit price. This also helps the reverse engineering team determine whether it met its goal of a 25 percent unit-cost reduction. Remember from the stage 1 report that the cost of the reverse engineered component is computed on the basis of a 25 percent unit-cost reduction.

In real-life reverse engineering, the stage 4 return on investment (ROI) can be lower than the 25:1 ratio used to select projects. The 25:1 ratio is used to maintain the integrity of the overall program even if individual projects have a lower ROI. Some projects will have a higher than 25:1 ROI, and this will help balance out the projects having a marginal ROI. Keeping in mind that any positive ROI is beneficial, the 25:1 ROI ratio is simply a desired goal.

As noted earlier, a unit-cost reduction of only 11 percent can be balanced by 42 percent unit-cost reduction from other projects. When the majority of the projects are reduced in cost less than the 25 percent goal, this might indicate a problem in either the prescreening criteria or in decisions from the early stages of reverse engineering. Initially selected projects should aim for the 25:1 goal, but as the program matures the average ROI is likely to decrease as the pool of high potential projects is depleted and those of a more marginal nature are introduced for reverse engineering. The approval body and financial persons will probably monitor the progress of the entire program on the basis of the team's ability to select projects wisely and achieve positive tangible economic results regardless of the 25:1 goal.

It is also helpful to compare the projected cost to complete stage 1 and those costs associated with the completion of stages 2 through 4 against the actual project costs to validate the estimating abilities of the projects which the team has undertaken. Some level of accuracy can mean that the technical staff know what they are doing and not just making guesses. Over time it will show which shops, laboratories, or other company resources are the most effective, or ineffective, as the case may be. (A company which has continuous improvement as its goal would then highlight this shop or lab for assistance in improving its effectiveness.)

Determination of the Return On Investment. This is the one piece of data certain to be watched closely. One objective of reverse engineering is to achieve a ROI of the organization's funds. To calculate the ROI, we must return to the calculations made for the stage 1 report. This figure was based on the target 25 percent unit-cost decrease. If the unit-cost decrease varies from the 25 percent, then the life-cycle savings (LCS) must be recalculated using the actual percentage cost reduction. The LCS calculation can be found in Chap. 3, on economic and logistics calculations. The ROI is a ratio of the total projected LCS savings minus the cost to reverse engineer (RE cost) (the actual money invested) divided by the cost to reverse engineer.

$$\text{ROI} = \frac{\text{LCS} - \text{RE cost}}{\text{RE cost}}$$

The LCS calculation may provide the largest variation in the actual stage 4 ROI. In the prescreen the decrease in unit cost is assumed to be 25 percent. By the time stage 4 is reached, there is a new estimate of the unit cost decrease due to reverse engineering. Thanks to our efforts in stage 3, we now know how much it should and does cost to produce this part. If we return to our previous example of a $537 part being purchasable for $372 after reverse engineering, this represents a 30.7 percent decrease in unit cost. This impacts the LCS when it is calculated as follows: (LCC = life-cycle cost)

$$\text{LCS} = (\text{LCC}) \, (\% \text{ unit savings})$$

where

$$\text{LCC} = (\text{life-cycle usage}) \, (\text{pre-RE unit cost})$$

If the part in question were needed 100 times per year for the remaining service life of 15 years, the life-cycle usage would equal $100 \times 15 = 1500$. Over the next 15 years 1500 parts are needed, assuming that the demand is held constant. With a pre-RE unit cost of \$537, the LCC would equal \$805,500. The LCS calculated in prescreen was

$$LCS = \$805,500 \times 25\% = \$201,375$$

The LCS calculated in stage 4 is

$$LCS = \$805,500 \times 30.7\% = \$247,288.50$$

This is an expected increase in actual savings of \$45,913.50! This is probably enough to cover the cost to reverse engineer the average project. This type of change from the projected to the actual costs or savings can prove significant and should be highlighted in the engineering and economic report.

Despite having a newer, more accurate set of figures with which to compute the ROIs and such, these figures remain relatively unproven until after the first buy using the reverse engineering technical data has been made. The stage 4 ROI is still a projected figure, while the stage 1 ROI was an estimate.

In many cases in which reverse engineering has been conducted, the engineers can point to intangible benefits gained from the process and its resultant new part. If there is any way to quantify the longer mean time between failure, which lowers overall maintenance costs, or the increased ease of installation, which shortens system downtime, it is recommended that an attempt to calculate these values be made. This information may turn the tide on a marginal part or provide some future gain as yet unknown.

The Final Technical Data Package Approval

This is the last leg of the journey for a part which has traversed many departments, many users, many individual contributions which now are collectively known as the *reverse engineering team*. Much human effort has been expended into making this a reality—not a cancerous, money-sucking black hole of paper. There may be cases where even under the best of efforts the approving body will not sign off on a project. If resolution can be reached, this would be a desired goal. Otherwise, resolution is unknown and the project can be effectively terminated (and no savings are achieved).

Despite the possibility of a last-minute snag, most projects which provide the organization with some tangible benefit will receive final approval. After all, who could turn down all that hard labor and design effort to make a prototype that works as well as, or better, than the part that was admitted into the reverse engineering program in the first place? It was already too expensive, too hard to purchase, and too few in existence; broke down too often; and/or did not work right in the first place. So rest assured, if the steps of each stage were executed to the best of the team's collective ability, virtually all reverse engineering projects will be approved.

The Approval Body

The final TDP now must be submitted for final approval. Exactly who makes up this final approving body is left to the discretion of the company, but there should be some overriding (or underwriting) activity to determine which parts will be manufactured and procured in accordance with the new technical data generated by the reverse engineering process. This approval committee is responsible for ensuring that the overall integrity of the system will not be jeopardized and that resources are not spent frivolously.

Why Final Approval Is Necessary

By now the reader must be wondering why final approval is necessary when there are approvals at each stage of the reverse engineering process. The multiple levels of approval are part of the checks and balances built into the process. Without the approvals and communication with the users and organizations which fund the program altogether, some large projects have not only failed but also totaled up large financial losses. If all the steps have been adhered to, the approval process should be only a formality. If the quality has been built into the end product at each step of the way, it need not be mandated to gain final approval.

One true-life project lost upwards of $250,000 of out-of-pocket company money because of inadequate screening, poor communication with the users, and ignoring the findings of stage 2 which pointed to project failure. The project was already over budget in stage 2, based upon the original estimate generated in the stage 1 report. The engineering expertise given in stage 2 to abort the project was overridden as the project went into stage 3. The project went to completion only to have the final reverse engineered part cost twice as much as the original. All the technical data generated ultimately were worthless to the agency which funded the project. The $250,000 spent could have been better utilized to fund additional future projects; however, there was no oversight function to step in and intervene when an elusive goal was being mindlessly pursued in a vain attempt to cover up errors in engineering judgment made early in the process.

The multiple checks and balances ensure that the program works to the benefit of those underwriting the team's efforts and prevents catastrophic loss. The oversight function is unnecessary if all participants have done their jobs.

Final Implementation

Once the project has gained signature approval, all future procurements will now represent the reverse engineered component configuration. The engineering drawings should now be signed, if they were not already signed at the completion of stage 3, by the lead engineer. Record copies should remain in engineering, while procurement should receive signed copies for future buys.

The shop floor personnel should be made aware that the reverse engineered design has been approved for final implementation, and any new part numbers should be added to the procurement files and the engineering database used for maintenance. With the inclusion of purchasing and procurement in stage 4, the shop floor personnel are now ready to fulfill their role in the execution of guiding future requestors to the new item when they receive requests for the old part. If a potential list of suppliers has been identified and generated, a notice should provide information on the date of final approval so that they can be prepared for any future requests for replacement parts. With the first purchase, implementation has begun.

8
Summary of Reverse Engineering

Classical examples of mechanical and electrical reverse engineering have been threaded throughout the text, but to highlight many important aspects of reverse engineering a single special project is reviewed in depth, step by step. *Special projects* are that elite class of reverse engineering that allow engineers to use all the creative faculties within their power. Design solutions are not the typical everyday part substitution; often they require the synthesis of much design information to produce workable solutions. Because of their inherent complexity, special projects often take long periods of time to complete and are very costly. Although the cost of special projects is high, substantial returns on investment stand in the balance. The return on investment based on a simple ratio increment of 40:1 can produce a high payback in real money.

Reverse Engineering Summarized in a Special Project

The example special project selected for review is a magnetometer used on the degaussing systems aboard surface ships. The magnetometer, a component of the degaussing system, senses the strength of the magnetic field in the X, Y, and Z directions. This project took almost 3 years to complete. The reverse engineering costs were estimated to be $250,000. The final cost was over $335,000. The projected return on investment was expected to be in the range of $800,000. With these high dollar amounts, the return on investment is only 1.4:1 expressed as a ratio; if this is realized, then the organization conducting the reverse engineering stands to save $465,000 over the remaining lifetime, and this is a substantial savings in any budget.

Prescreening and Stage 1

When the magnetometer was reviewed in the prescreen, it was an $11,000 nonrepairable item of supply. Because of the complexity of the design, high failure rates were being experienced. If this part was installed and failed to work for any reason, there was no way to repair or adjust the magnetometer, and another $11,000 part had to be requisitioned from supply. When the lid was removed, the magnetometer revealed all the internal components solidly entrenched on a bed of epoxy resin that could not be removed without damaging these internal components. This served to further reinforce the magnetometer's nonrepairable status by making it a throwaway product in the event of failure.

Despite the formidable challenge of reverse engineering a design as yet unaccessible, the economics of the project reinforced the need to proceed with reverse engineering. The remaining service life was 20 years because this component was being installed in a class of ships still under construction for the U.S. Navy. The annual usage rate was 14 to 15 units per year. This means that approximately 290 units would be needed over the lifetime of this part. At $11,000 per unit, that equates to a life-cycle cost of $3.19 million. A unit savings of 25 percent equals a life-cycle savings of $797,500. In stage 1 the original cost to reverse engineer was estimated to be $250,000. The magnetometer was classified as a data development project. With these figures in hand, it was decided that this was an economical reverse engineering project. Because of its complexity and the associated risk, the high cost, and the possibility that this could take 2 years to complete, this project was classified as a special project.

Stages 2 through 4

The first challenge to reverse engineering the magnetometer was to get the internal components out of the casing. As mentioned earlier, all the internal components were solidly encased in epoxy resin. The first line of attack was to physically chip out the epoxy. This method caused damage to the first components encountered and was abandoned. A second method using powerful solvents created little further progress. At this point a fortuitous opportunity befell the engineering staff. A small business was demonstrating its proprietary method of depotting solid epoxy resins without damaging the parts encased in the resin. After examining the magnetometer to determine the feasibility of depotting this unit, this small company decided that, indeed, they could unentomb the contents of the magnetometer. The part was shipped off and a few weeks later returned with all its internal components undamaged and intact. The results of this process not only left the internal components undisturbed but also the part numbers still legible (a tremendous aid in the reverse engineering process). It is regrettable that this small business no longer exists and the proprietary process remains a secret to this day.

Figure 8-1 shows the layout of the internal components and their relative sizes after the depotting process. Only one of the three magnetic cores is shown. The design then had to be studied in great detail, and a draft schematic of the now depotted magnetometer was drawn. This schematic was needed to begin to understand the original design strengths and weaknesses. Because there was no damage to the internal components, the defects were repaired and operational bench testing was performed to determine many of the operational characteristics.

Much of the design utilized 1970s electronics technology; however, there were some unique and relatively ingenious applications of this technology. The most difficult design parameter to reverse engineer was the nine interdependent potentiometers which had to be adjusted perfectly to a cali-

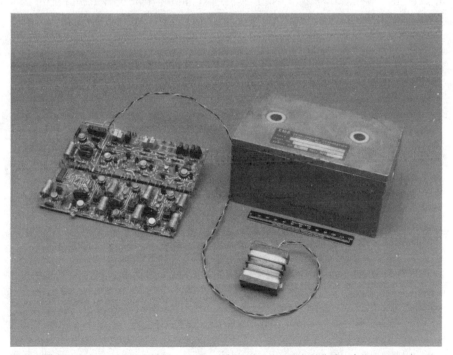

Figure 8-1. Layout of internal components of original design and their relative sizes after the depotting process.

brated core value. Each and every magnetometer had been calibrated for the individual set of cores installed in the unit. Once all nine potentiometers of the unit were adjusted to match a calibrated set of core values, the unit was installed into the aluminum casing and filled with epoxy to keep it from being disturbed and knocked out of calibration.

EMI Shielding. A unit operating in the field was exposed to high electromagnetic interference (EMI) fields and therefore required considerable EMI shielding. This unit was also exposed to an open saltwater environment and on the basis of ship missions could be required to operate in temperatures ranging from those experienced along the equator to those experienced within the arctic circle. The magnetometer had to continue accurate operations in a very difficult environment, and so filling the casing with epoxy was not an undesirable design solution, unless you happen to be reverse engineering this unit. Potting all the internal components with epoxy, however, is not the only solution to meeting these strenuous design criteria.

One of the intrinsic goals of reverse engineering is to reduce the unit cost by 25 percent. The magnetometer project also had a secondary goal: to make this a repairable item using a set of skilled individuals to maintain this item. Repairable items are available in a variety of forms: those that can be repaired only by the original manufacturer, those that can be repaired at one single location or depot, those that can be repaired only on shore in the case of the U.S. Navy, and those that can be repaired by any skilled technician in the field, or at sea in this case. The magnetometer had a reverse engineering philosophy that repair and recalibration could be performed at a single depot to have them available for reinstallation in the field in the event a unit failed.

To render the magnetometer repairable, another method of encasing the internal components, in lieu of the epoxy, had to be found. After reviewing the field of alternatives, a cast aluminum casing used for cable TV line amplifiers was found. This casing had the high EMI shielding already available because of the Federal Communications Commission (FCC) strict radio frequency signal leakage specifications. These casings also have the capability to withstand cyclic extremes of temperature and moisture for years on telephone poles and other unprotected areas. This type of casing could then provide the level of EMI and environmental protection for a much lower cost since these were already being mass-produced for the TV cable industry. The only difference that posed a problem was the larger size of the cable TV encasement. Due to restrictions of the mounting frame and size, a call had to be placed to a knowledgeable person who could determine whether a larger casing could be accommodated. A field user was identified and contacted who did verify that the larger size could be used but that the bolting pattern had to be maintained for interchangeability

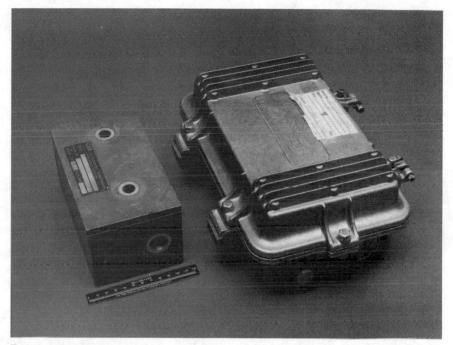

Figure 8-2. Size difference between original and reverse-engineered casings.

between the original equipment and the reverse-engineered model. Figure 8-2 shows the physical size difference between the two casings.

The Magnetic Cores.   With the EMI shielding issue settled, the next design challenge was to reverse engineer the magnetic cores. The magnetic cores in the original design were no longer available on the commercial market. Similar cores were identified, but a new method of wrapping the cores with copper windings to give them the same essential characteristics was needed. The multilayer construction of the newer cores is shown in Fig. 8-3. It is also apparent that these new cores are approximately one-third the size of the older cores after being wound in copper.

The Breadboard Model.   With the original design committed to memory with all its flaws and ingenious solutions, the path for the development of new design circuitry had to be mapped out. The method which was opted for to test out new electrical circuit ideas was the breadboard model. The breadboard model minus the X, Y, and Z cores is shown in Fig. 8-4. The use of the breadboard allowed for testing of newer design options than were previously available in the 1970s when this design was originated. Many of the older components were replaced by integrated-circuit (IC) chips de-

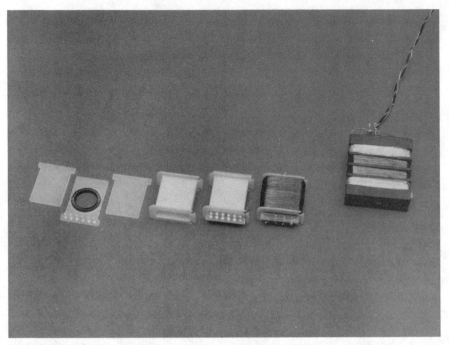

Figure 8-3. Multilayer construction of the reverse engineered cores.

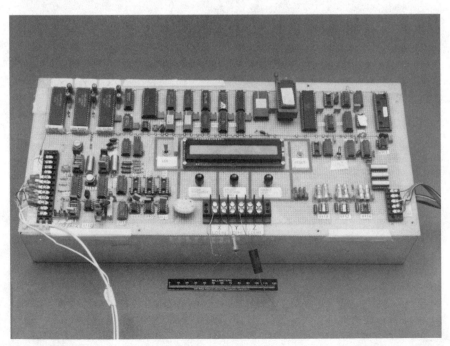

Figure 8-4. Breadboard model minus *X*, *Y*, and *Z* cores.

signed to achieve specific performance characteristics. The use of IC chips improved the reliability of the individual components and of the overall unit. The nine interdependent potentiometers were replaced with a microprocessor chip that could solve a calibration matrix for the nine unique coefficients. Figure 8-5 shows the breadboard model with the magnetic cores embedded in a test assembly to verify that measurements of the magnetic field are accurate in all directions and with all degrees of freedom.

Old versus New Circuitry Design. Figure 8-6 shows the differences in circuitry technology between the older design and the newer design, which has built-in diagnostic and test capabilities. The upgrade from a two-circuit-card assembly (left) to a five-circuit-card assembly (right) allowed for the use of an extender card to be used as a base to attach a circuit test and diagnostic capability. The accuracy of measurements on each axis could be tested and in situ recalibration could be accomplished. Figure 8-7 shows the entire assembly of internal components necessary for the reverse engineered design: casing, cores, and circuitry. Figure 8-8 shows the assembly as it is installed in the casing with the test card assembled onto the extender card.

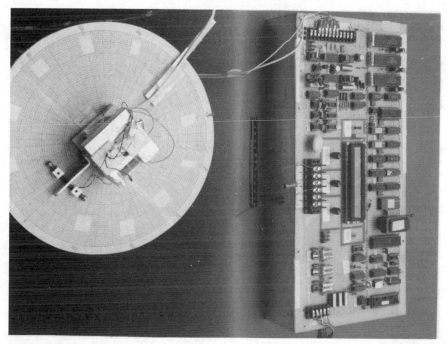

Figure 8-5. Breadboard model with cores.

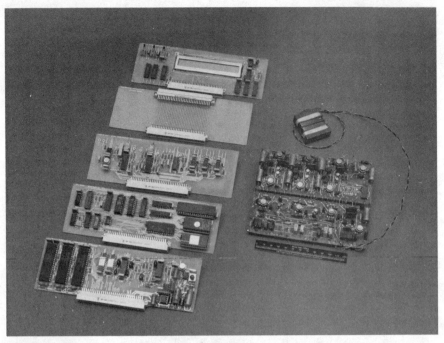

Figure 8-6. Circuitry technology differences.

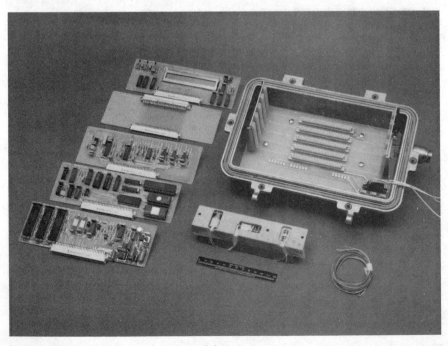

Figure 8-7. Layout of reverse engineered design.

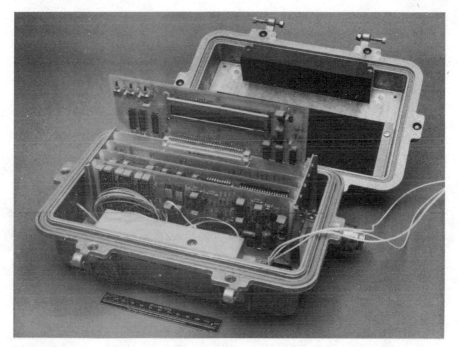

Figure 8-8. Assembled reverse engineered design.

Review of the Project as a Successful Reverse Engineering Model

Only one prototype of the new magnetometer design was produced. The breadboard model was put through rigorous testing before committing the design to a prototype model. The prototype model was later bench-tested equally rigorously and system testing was to begin. The prototype magnetometer was estimated to cost around $8000, representing a 27 percent unit-cost savings. It also achieved the repairability level that was expected. The final outcome of this project is indeterminate due to a lack of commitment on the part of the user.

The review of this project does serve to highlight the multitude of steps and resources necessary for successful reverse engineering. The casing dilemma served to highlight the interdisciplinary talents required for many projects. The IC chips are an example of product substitution. The synthesis of the new design is the product of many talented minds working toward a common goal and, all restraining factors withstanding, achieved that end from an engineering standpoint.

9
Future
Applications

Where does reverse engineering go from here? How does it fit into today's changing global technological and economic environment?

Solidly entrenched in the mid-1990s, reverse engineering seems to be experiencing a revival with the United States Department of Defense (DoD), along with a growing interest in the subject in the commercial arena. In the DoD it seems only natural that in these times of tight military budgets that current systems need to be maintained longer and longer. Most military systems are designed with a 30-year life span. The current rapid pace of technological growth has served to highlight reverse engineering as one of the more economical solutions to the diminishing sources of manufacture and supply. What the military has felt for over a decade the commercial arena has just now begun to experience. The use of IC chips in designs and their typical life cycle of 3 to 5 years has led to the increasing awareness in industry of the problem the military has faced for some time. Figure 9-1 shows the relationship between the volume of use of a particular IC chip and its life cycle for the industrial and military markets over time.

The future always looks different from our vision. Invariably we cannot see the future, only the trends, changes, or challenges facing us today. Some trends do have a tendency to persist in appearing in our everyday world already, although we may not be ready to accept the challenges of accommodating these ideas. Ideas such as corporate downsizing and right-sizing are now acceptable, but accommodating this idea means requiring more effective use of current capabilities with higher reliability using a smaller workforce which is required to maintain or improve productivity. Using yesterday's cumulative knowledge, we cannot see that this possibility exists because this multitude of requirements appear to be contradictory

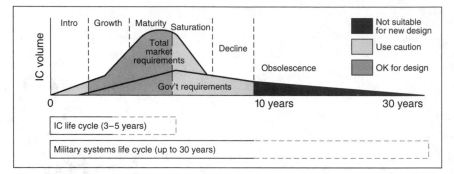

Figure 9-1. Integrated-circuit production volume over time.

and therefore impossible to satisfy. These are the future requirements, and only those who can synthesize creative new applications with quantum leaps in knowledge will find themselves a competitive force in the future; the rest will be playing catch-up.

To achieve these quantum leaps the use of "technology" has been the all-too-easily thrown gauntlet; however, the use of technology for its own sake has often been thrown out by those who need it most as an effective and affective proactive solution. Some industries have even thrown the baby out with the bathwater looking to no new technology or technological investment to provide them a competitive edge. Instead they simply want the worker to work harder, be more productive, or take fewer breaks.

The use of improved tools to do these jobs is as important as the additional training, or retraining, in the use of these tools. The admirable purpose of business process re-engineering is aimed at reevaluating the entire process or system at hand before automating antiquated methods of production, whether it be the production of automobile parts or the processing of paperwork, or even chickens through chicken processing.

While improved tools and training cost money, improved communication between departments and divisions does not. One positive and unfunded aspect of reverse engineering is the need for multidisciplinary teams to work together, many for the first time. This opens communication between not only engineering disciplines but also divisions such as engineering and the shops and/or production-line personnel. There has been a tendency for the formation of caste systems or socioeconomic hierarchies between engineers and technicians or equipment operators. Communication between the various types of skilled personnel can be improved on when they all work as a team to achieve an objective, thereby increasing the respect each individual gives to the others. On the negative side, this communication can serve to point to ineffective engineers, overcompetent technicians who should be sent to get engineering degrees, or a decrease in the number of technical skills needed to operate equipment. In short, it probably means change, both for the actual workforce and for the way management thinks of its personnel. Reverse engineering is a first step to opening the door of communications between functional areas; however, there is a multitude of options beyond this to improve communication between people on various portions of any given process.

Electronic mail (e-mail) systems which allow people to pose detailed technical questions without disturbing day-to-day workflow operations offer the capability of communications at the convenience of the people it is to serve. Communications of convenience are nonintrusive and effective for non-time-critical discussions of a less than detailed or interactive nature. People who must work together form natural networks. Who in the workplace does not have a natural grouping of individuals one can call on to assist in getting a task accomplished? How could any of us achieve our daily duties without the assistance of a natural functional network? In order to utilize e-mail, a facility must have computers and local area networks. For those who are familiar with the internet, this may seem incredibly basic, but there are many industries from shipbuilding to publishing which do not have wholesale access to interactive networks. Beyond access to the equipment, there needs to be a reason to use this form of communication. The lead engineer would have the most influence over the full implementation and use of an electronic networking capability, which will decrease decision-making time, thereby speeding up the whole reverse engineering process.

From this point concerning interactive real-time communications capabilities, it is a short leap into concurrent reverse engineering. Concurrent reverse engineering would shorten the entire process time from the time the part enters stage 1 to the time it completes stage 4. Chapters 4 through 7 detail in a step-by-step manner the manual method of developing technical data. In the future many of these manual methods will take advantage of effective automated and/or combined technologies which rely on today's integrated software programs. In concurrent engineering today, the digital data which are developed while measuring the surface features during the dimensional inspection can also be used for milling, inspection, and documentation processes later. If the manual methods are shortened in the reverse engineering time cycle through the use of automation, there can be considerable reverse engineering program cost savings as a result of this data integration. The decrease in the time necessary to complete all stages from prescreening to prototype testing shortens the time to realize cost savings.

The future of efficient reverse engineering will certainly include automated methods of reverse engineering using integrated systems to measure, manufacture, inspect, and test components. Regional centers who have the equipment to handle this operation would certainly receive business from the local communities in many industries, from radio to electronics to handmade furniture items. In some cases just the development of detailed and accurate technical data would suffice. In other instances, multiple small- and medium-size lots would ensure the operation of a capability until new systems could be installed to replace those with obsolete components. Many manufacturers of parts similar to obsolete components have been approached to reverse engineer a similar part. Since there was little demand (i.e., little profit), the needed part was turned away because the tooling costs were too prohibitive. With an increasing demand for one-of-a-kind items and a flexible manufacturing environment that could be adaptive or agile depending on customer needs, this reverse engineering business could thrive as a niche operation.

A Systems Approach

From individual parts we now turn to a systems approach. If a completed reverse engineered part is more reliable than the older design, there are associated increases in system reliability, shortened system downtime, and higher output productivity levels. This can result from the successful implementation of merely one reverse engineered part. If one studied an entire system for reverse engineering-critical components, one could extrapolate the effect on the overall system if 10 parts were completed and implemented at once. These 10 parts can be reverse engineered virtually simultaneously; therefore, interference fit can be considered to be a less critical issue than in stand-alone components. If the entire data set is digitized, then all interrelated parts can be put in three-dimensional format, allowing much operational information to be verified before prototype testing in an operational system. This is a form of process synthesis. Synthesis of information is the area of reverse engineering which can achieve the greatest gains and process improvements. The synthesis of design elements and requirements for multiples of related parts can have a tremendous leverage in increasing the ROI when combined. What if one circuit card were reverse engineered because of its high failure rate in a black box which contained 10 similar cards? The knowledge gained about the first reverse engineered part could be leveraged to decrease the cost of the remaining nine circuit cards. Even if they had a lower failure rate and lower return on investment, would not the knowledge of the remaining nine functional characteristics be inexpensive to obtain? If these nine circuit cards are part of an integral system and these were reverse engineered also, would not the 10 circuit cards combined have a greater effect than the one alone? In certain cases of reverse engineering (and life) the sum of the whole is greater than the sum of the parts—and this is true synthesis.

Concluding Remarks

One primary idea pervades these ideas: the digitization of all our work. Our drawings are digitized, our information is exchanged digitally, our identities are digits on forms with our names reduced to on or off bits. The ASME standard Y14.26M—1989, *Digital Representation for Communication of Product Definition Data,* reprinted in part as Appendix C, hits on more than one major theme. That our work will be translated into a digitized format is one idea. That we need to communicate this information in a common fashion is another idea, and thirdly, that our data will be represented by these digits and not be a stand-alone reality.

Most of our future work will be modelled and tested without ever touching the base material(s) until all analysis and testing has been conducted on a digitized model. CAD/CAM was just the beginning. Real-time modeling and concurrent reverse engineering are a marvelous synthesis of multiple technologies that have little value if taken separately, but produce a great effect on the operation of reverse engineering if combined into a coherent package. This is the type of informational leap we are searching for when system improvements are needed.

In many companies there is too much duplication of effort from department to department. Leveraging design knowledge and facilities usage to effectively solve more than one piece of one problem is the goal of effective information processing. If the Department of Energy has a waste management facility which has conducted extensive research in my area of interest as a manufacturer of a product that has a toxic by-product from the manufacturing process, why should I reinvent the waste management strategy? If, during a collective investigation into the technology area it is found that no other group has research to address my need, I can become the leader in that technological area. If it has been done before, borrow as much information as necessary to solve the problem.

No one can tell a design engineer how to do their job, one can only explain the task that needs to be accomplished and the steps involved in the accomplishment of this task. The devices chosen by the competent skilled worker to do this job are a measure of creativity and skill. What could be more rewarding to an engineer than creating a device, a machine of harmony, simplicity, and beauty of design. Learning from the engineers that either came before us or are our peers can teach us about their brilliance and ingenuity, and we can then use this knowledge to create the designs of the future. Accessing the information we need to solve our engineering problems is the true trick of the trade. In today's environment we need each other to solve today's problems. Can anyone imagine that we will need each other less in the future? As Albert Einstein stated, "If I have seen farther it is because I have stood on the shoulders of giants." In reverse engineering there is the ability to unlock design secrets and develop innovative solutions to older design problems, or problems created by older designs in newer systems.

Reverse engineering can be likened to the act of communication between a person using an old 286 personal computer from an UNIX workstation user via a communications platform that allows the 286 machine equal footing as the UNIX workstation, which in turn allows both people to exchange ideas and be productive. So it is with successful reverse engineering, finding common engineering communications to keep the 286s of the world productive. Continuing productivity with existing equipment is still paramount to survival and success. It may be nice if the world market were a level playing field, but it is not; yet we must play in the fields we are given. Reverse engineering helps those who have older playing fields compete with those who have bright and shiny equipment in a brave new world by sharing and developing detailed technical information.

Appendixes

Throughout the text there have been many references to standards. These engineering drawing standards are very important to the consistent application of drawing practices and the interpretation of drawing symbols to accurately recreate the product originally conveyed by the design engineer. Standardization will allow for the largest possible audience to understand the design across all cultural and language barriers. The promulgation of and the use of these basic standards within all organizations is encouraged for the benefit of international trade and economic growth in the twenty-first century.

The excerpts of the standards presented in the following Appendixes are particularly useful in that they form the building blocks of most engineering drawings. The table of contents from each is included to demonstrate the scope of each document. The foreword is included to give a sense of the intent of the document. The opening pages of the standard detail the scope and applicable associated documents. Copies of the complete document can be obtained from either engineering or public libraries or from the organization that is charged with maintaining the document.

Appendix A presents excerpts from ASME Y14.24M-89, Types and Applications of Engineering Drawings. This standard lists all acceptable engineering drawing variations in all levels of detail. While most of the documentation practices are slanted toward mechanical components, there is a section on electrical/electronic drawings that was briefly mentioned in Chap. 5. For those dedicated to the reverse engineering of electrical/electronic parts there will be additional requirements, too many to try to cover in this excerpt.

Appendix B reprints excerpts from ASME Y14.34M-89, Parts Lists, Data Lists, and Index Lists. This standard is meant to be a companion document to ASME Y14.24 since all drawings will have parts listings. Guidance on data lists and index lists are provided for those instances where parts lists are not suitable to best describe the item on hand.

Appendix C reprints excerpts from ASME Y14.26M-89, Digital Representation for Communication of Product Definition Data. In the years ahead this may become the most important of these standards for design engineers. Most design work is now being done on engineering workstations. The software associated with computer-aided design and manufacture will need to be interpreted in an identical manner by all software programs, whether it is for the programming of numerically controlled equipment, laser coordinate measuring systems, or automated inspection equipment. From a review of the table of contents it is obvious that not all engineers will actually use this standard in their everyday work; however, it is important to be aware that all equipment must be interoperable and in compliance with this standard.

Appendix D presents excerpts from ASME Y14.5, Dimensioning and Tolerancing. Along with Y14.24 and Y14.34 these three documents form the building blocks of engineering drawing practice. Dimensioning and tolerancing often mean the difference between a quality part and an unacceptable one. Contractual acceptance of parts often depends on proper dimensioning and tolerancing. This standard is critical to the production and quality of parts.

Throughout the text other documents are noted but not expanded upon. This is due largely to their relative application. While the military standards and specifications are critical to anyone in the defense or government acquisition and supply business, these are not globally accepted standard practices. If they are to take precedence to industrial standards the user is probably already aware of the contents of those documents. The specifications and standards listed below are not meant to be comprehensive; they are meant to highlight some other documents of guidance to design and engineering practice.

MIL-STD-100E.   Engineering Drawing Practices

MIL-T-31000.   General Specification for Technical Data Packages

MIL-Q-9858.   Quality Program Requirements

MIL-I-45208.   Inspection System Requirements

ANSI Y14.15.   Electrical and Electronic Diagrams

ANSI/IEEE STD 991.   Preparation of Logic Circuit Diagrams

Appendix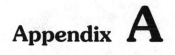

AN AMERICAN NATIONAL STANDARD

ENGINEERING DRAWING AND RELATED DOCUMENTATION PRACTICES

Types and Applications
of Engineering Drawings

ASME Y14.24M-1989

The American Society of
Mechanical Engineers

345 East 47th Street, New York, N.Y. 10017

FOREWORD

(This Foreword is not part of ASME Y14.24M-1989.)

This Standard was prepared to define the accepted drawing types used to establish engineering requirements. Each type is defined by general description, application guidelines, and specific content requirements. Work on this Standard considered the types of *engineering drawings* most frequently used by business, industry, and government communities in the United States of America in the production and procurement of hardware. This Standard attempts to serve the individual and combined needs of these communities and assure consistency of application and interpretation.

Drawing definitions are intended to permit preparation by any suitable method (manual, computer-aided, photographic, etc.); therefore preparation techniques or methods of reproduction are not described.

On these bases, a series of meetings were held to identify, select, and prepare proposed text and illustrations. At each stage of development, the Subcommittee considered the selection of elements best suited for a national standard. Members of the Y14.24 Subcommittee represented a cross section of American industry and the Department of Defense (DoD). Liaison with technical societies such as the American Defense Preparedness Association (ADPA), Electronic Industries of America (EIA), and Society of Automotive Engineers (SAE), provided additional technical support.

In addition to the current Subcommittee Members who helped finalize this Standard, the following personnel served in the development of its content:

O. C. Baker, Texas Instruments Inc.
D. S. Bennett, Jr., Air Force Logistics Command
P. D. Calvin, TRW Defense & Space Systems Group
M. S. Clifton, U.S. Naval Ship Missile Systems Engineering Station
L. W. Cornell, Boeing Aerospace Co.
D. L. Edmunds, Martin Marietta Corp., Data Systems Division
G. Garrard, U.S. Naval Sea Systems Command
R. A. Graefe, McDonnell Douglas Corp.
J. Hobko, U.S. Army Electronics Command
C. A. Nazian, U.S. Army Munitions Command
T. C. Pritchard, Lockheed Missile and Space Co.
M. Randolph, United Airlines
A. D. Signor, U.S. Naval Ship Weapon Systems
M. E. Taylor, U.S. Army Armament Research, Development, and Engineering Center
J. U. Teague, Lockheed California Co.
R. A. Timlin, Martin Marietta Corp.
J. L. Zeno, Air Force Logistics Command

Commendation is extended to the companies and DoD Departments and Agencies for sponsoring participants in this activity and to those whose earlier efforts provided the basis for the types of drawings in this Standard. The success of this effort can be attributed to their demonstrated interest, cooperation, and support.

Coordination of this Standard with the International Standards Organization (ISO/TC10/SC1) is intended to help enhance world understanding of the various types of drawings in use within the United States.

Suggestions for improvement of this Standard are welcomed. They should be sent to the American Society of Mechanical Engineers, Standards Department, 345 East 47th Street, New York, NY 10017.

After approval by the Y14 Committee and the sponsor, and after public review, this Standard was approved as an American National Standard by the American National Standards Institute, Inc., on November 3, 1989.

CONTENTS

Appendices

Figure

Tables

ASME Y14.24M-1989

ENGINEERING DRAWING AND RELATED DOCUMENTATION PRACTICES

TYPES AND APPLICATIONS OF ENGINEERING DRAWINGS

1 GENERAL

1.1 Scope

This Standard defines the types of *engineering drawings* most frequently used to establish engineering requirements. It describes typical applications and minimum content requirements. Drawings for specialized engineering disciplines (e.g., marine, civil, construction, optics, etc.) are not included in this Standard.

1.2 Applicable Documents

When any American National Standard listed in the following paragraphs is superseded by a revision approved by the American National Standards Institute, Inc., the revision shall apply to the extent specified herein.

1.2.1 Documents Referenced in Text. The following documents form a part of this Standard to the extent specified herein. Unless otherwise specified, the latest issues of the documents apply.

American National Standards

ANSI/IEEE STD 91-1984, Graphic Symbols for Logic Circuit Diagrams

ANSI/IEEE STD 200-1975 (R1988), Reference Designations for Electrical and Electronics Parts and Equipments

ANSI/IEEE STD 315-1975 (R1988), Graphic Symbols for Electrical and Electronics Diagrams

ANSI/IEEE STD 991-1986, Preparation of Logic Circuit Diagrams

ANSI Y14.15-1966 (R1988), Electrical and Electronic Diagrams

ASME Y14.34M-1989, Parts Lists, Data Lists, and Index Lists

Government Document

Military Handbook II4/II8, Commercial and Government Entity Codes

1.2.2 Other Applicable References

American National Standards

ANSI/AWS A2.4-1986, Graphic Symbols for Welding, Brazing, and Nondestructive Examination

ANSI/IEEE STD 260-1978 (R1985), Standard Letter Symbols for Units of Measurement (SI Units, Customary Inch-Pound Units, and Certain Other Units)

ANSI/IEEE STD 268-1982, Standard Metric Practice

ANSI STD 280-1985, Standard Letter Symbols for Quantities Used in Electrical Science and Electrical Engineering

ANSI/IEEE STD 623-1976, Graphic Symbols for Grid and Mapping Used in Cable Television Systems

ANSI/IPC-D-350-1978 (R1987), Printed Board Description in Digital Form

ANSI/IPC-T-50D-1988, Terms and Definitions for Interconnection and Packaging Electronic Circuits

ANSI/ISA S5.1-1984, Instrumentation Symbols and Identification

ASME Y1.1-1990, Abbreviations for Use on Drawings and in Text

ANSI Y10.1-1972 (R1988), Glossary of Terms Concerning Letter Symbols

ANSI Y10.3M-1984, Letter Symbols for Mechanics and Time Related Phenomena

ANSI Y10.4-1982 (R1988), Letter Symbols for Heat and Thermodynamics

ANSI Y10.10-1953 (R1973), Letter Symbols for Meterology

ANSI Y10.11-1984, Letter Symbols and Abbreviations for Quantities Used in Acoustics

ANSI Y10.12-1955 (R1988), Letter Symbols for Chemical Engineering

ANSI Y10.17-1961 (R1988), Greek Symbols for Math

ANSI Y10.18-1967 (R1987), Letter Symbols for Illuminating Engineering

ANSI Y14.1-1980 (R1987), Drawing Sheet Size and Format

ANSI Y14.2M-1979 (R1987), Line Conventions and Lettering

ANSI Y14.3-1975 (R1987), Multi and Sectional View Drawings

ASME Y14.4M-1989, Pictorial Drawings

ANSI Y14.5M-1982 (R1988), Dimensioning and Tolerancing

ANSI Y14.6-1978 (R1987), Screw Thread Representation

ANSI Y14.7.1-1971 (R1988), Gear Drawing Standards - Part 1 for Spur, Helical, Double Helical, and Rack

ANSI Y14.7.2-1978 (R1989), Gear and Spline Drawing Standards - Part 2, Bevel and Hyphoid Gears

ASME Y14.8M-1989, Castings and Forgings

ANSI Y14.13M-1981 (R1987), Mechanical Spring Representation

ASME/ANSI Y14.18M-1986, Drawings for Optical Parts

ASME Y14.26M-1989, Digital Representation for Communication of Product Definition Data

ANSI Y14.36-1978 (R1987), Surface Texture Symbols

ANSI Y32.10-1967 (R1987), Graphic Symbols for Fluid Power Diagrams

ANSI Y32.11-1961 (R1985), Graphic Symbols for Process Flow Diagrams in the Petroleum and Chemical Industries

ANSI Y32.18-1972 (R1985), Graphic Symbols for Mechanical and Acoustical Elements as Used in Schematic Diagrams

Government Document

MIL-STD-12, Abbreviations For Use On Drawings, and in Specifications, Standards, and Technical Documents

DOD-STD-100, Engineering Drawing Practices

1.2.3 Order of Precedence. In the event of conflict between the text of this document and the references cited herein, the text of this document takes precedence.

1.2.4 Source of Documents. The documents listed may be obtained as follows:

(a) ANMC: Application for copies should be addressed to the American National Metric Council, 1010 Vermont Avenue NW, Washington, DC 20005.

(b) ANSI: Application for copies should be addressed to the American National Standards Institute Inc., 1430 Broadway, New York, NY 10018.

(c) ASME: Application for copies should be addressed to the American Society of Mechanical Engineers, 345 East 47th Street, New York, NY 10017.

(d) IEEE: Application for copies should be addressed to the Institute of Electrical and Electronic Engineers, Inc., 345 East 47th Street, New York, NY 10017.

(e) IPC: Application for copies should be addressed to the Institute for Interconnecting and Packaging Electronic Circuits, 7380 North Lincoln Avenue, Lincolnwood, IL 60646.

(f) Government Documents: Copies are available from the Standardization Documents Order Desk, Bldg. 4D, 700 Robbins Avenue, Philadelphia, PA 19111-5094.

1.3 Definitions

Definitions of italicized terms used in this Standard are contained in Appendix A.

1.4 Methods and Styles of Preparation

Preparation methods (manual, computer-aided, photographic, "cut and paste," etc.), method of depiction (orthographic, pictorial, or exploded views), and styles (multisheet, book form, computer printout, etc.) are a concern of this Standard only to the extent that the drawing satisfies its intended purpose.

1.5 Illustrations

Sample drawings and other illustrations are included as needed to illustrate the text and the characteristics unique to a particular drawing type. To comply with the requirements of this Standard, actual drawings shall meet the content requirements and application guidelines set forth in the text. The content and arrangement of sample drawing types are for illustration only.

1.6 Application Guidelines

Application guidelines are intended to aid in understanding the conditions under which specific types of drawings may be prepared. It is not intended that any application guideline imply that preparation of specific drawing types is always required.

1.7 Drawing Content

Requirements may be satisfied by direct delineation on the drawing or by reference to other documents which are a part of the drawing package. Such documents are invoked in individual drawings either by general notes, in the using assembly parts list, or both. Parts lists shall be in accordance with ASME Y14.34M.

1.8 Tabulation

Any drawing type may be tabulated to delineate similar *items* which, as a group, have some common characteristics and some variable features.

1.8.1 Application Guidelines. Tabulated drawings are prepared to avoid preparation of individual drawings for each similar item tabulated.

1.8.2 Requirements. The differences (variables between the items) defined by the drawing are tabulated. The common characteristics are delineated or stated once. Each item is uniquely identified. Normally, a single pictorial representation is shown. For example: variable dimensions are coded by letters used as headings for columns in a tabulation block. Variables are entered in the table under the appropriate heading and on the same line as the *unique identifier* for the specific item. The description for each tabulated item is as complete as that of an individual item described on the specific drawing type.

1.9 Combination of Drawing Types

The characteristics of more than one drawing type may be combined into a single drawing provided the resulting combination includes the data required by the individual types. For example: a detail assembly drawing combines the detail of an item(s) and the assembly of which it is a part; a modification kit drawing combines a description of the modification and the kit of items needed to accomplish the modification.

NOTE: While combining is permitted, the decision to combine drawing types should be made cautiously. Some significant benefit(s) should out weigh such potential disadvantages as: (1) increased complexity which may diminish clarity and usefulness and (2) accelerate change activity of the combined drawing which may increase the need to update associated records, material control data, manufacturing planning, microfilm, etc.

1.10 Ancillary Drawings

Ancillary drawings may be prepared to supplement *end product* drawings. Ancillary drawings may be required for management control, logistics purposes, configuration management, and other similar functions unique to a *design activity*. Inclusion of data in an ancillary drawing does not eliminate the need to prepare appropriate drawing types, including the applicable data as defined in this Standard.

2 LAYOUT DRAWING (Fig. 4)

2.1 Description

A layout drawing depicts design development requirements. It is similar to a detail, assembly, or installation drawing, except that it presents pictorial, notational, or dimensional data to the extent necessary to convey the design solution used in preparing other engineering drawings. Except as specified in 2.3(k), a layout drawing does not establish *item identification*.

2.2 Application Guidelines

A layout drawing may be prepared for a complete end product or any portion thereof and is prepared either as:

(a) a conceptual design layout to present one or more solutions for meeting the basic design parameters and to provide a basis for evaluation and selection of an optimum design approach;

(b) a design approval layout to present sufficient details of the design approach for cost estimating and design approval;

(c) a detailed design layout depicting the final development of the design in sufficient detail to facilitate preparation of detail and assembly drawings;

(d) a geometric study to develop movement of mechanical linkages, clearances, or arrangements.

A layout is not normally used to fabricate equipment; however, a detailed design layout is sometimes used as an interim assembly drawing for development equipment.

2.3 Requirements

A layout drawing includes, *as applicable*:

(a) location of primary components;

(b) interface and envelope dimensions including a cross-reference to applicable interface control documentation;

(c) paths of motion;

(d) operating positions;

(e) critical fits and alignments;

(f) selected materials, finishes, and processes;

(g) wire bundle, pneumatic line, and hydraulic line routing and sizes;

(h) adjustments;

(i) critical assembly details and sequence;

(j) identification for selected *purchased items* and new design items;

(k) identification for the *assembly* depicted (when the layout is to be used as an interim assembly drawing).

A layout is drawn to scale with sufficient accuracy and completeness for its intended use.

3 DETAIL DRAWING

A detail drawing provides the complete end-product definition of the *part* or *parts* depicted on the drawing. A detail drawing establishes item identification for each part depicted thereon.

3.1 Monodetail Drawing (Figs. 5 and 6)

3.1.1 Description. A monodetail drawing delineates a single part.

NOTE: A drawing detailing SHOWN and OPPOSITE parts using a single set of views is considered to be a tabulated monodetail drawing (see 1.8).

3.1.2 Application Guidelines. A monodetail drawing is prepared to provide maximum clarity in defining a part.

3.1.3 Requirements. A monodetail drawing delineates all features of the part including, as applicable: configuration, dimensions, tolerances, materials, mandatory processes, surface texture, protective finishes and coatings, and markings.

3.2 Multidetail Drawing (Fig. 7)

3.2.1 Description. A multidetail drawing delineates two or more uniquely identified parts in separate views or in separate sets of views on the same drawing.

3.2.2 Application Guidelines. A multidetail drawing is a single drawing prepared to describe

parts usually related to one another.

NOTE: The decision to use a multidetail drawing should be made cautiously. The same revision status applies to all details on a multidetail drawing; therefore a change to one detail of the drawing may affect the associated records of all other details (material control data, manufacturing planning, microfilm, etc.). Some significant benefit(s) should outweigh this potential disadvantage, as well as such others as diminished clarity and usefulness resulting from increased drawing complexity.

3.2.3 Requirements. Each part delineation on a multidetail drawing meets the design definition requirements for a monodetail drawing and shall be uniquely identified.

4 ASSEMBLY DRAWING (Figs. 8 and 9)

4.1 Description

An assembly drawing defines the configuration and contents of the assembly or assemblies depicted thereon. It establishes item identification for each assembly. Where an assembly drawing contains detailed requirements for one or more parts used in the assembly, it is a detail assembly drawing (see 1.9 and Fig. 9).

4.2 Application Guidelines

An assembly drawing is prepared for each group of items that are to be joined to form an assembly and that reflect one or more of the following:

(a) a logical level in the assembly or disassembly sequence;

(b) a testable item;

(c) a functional item;

(d) a deliverable item.

A detail assembly drawing is the preferred drawing type for an *inseparable assembly*; however, individual pieces of an inseparable assembly need not be individually detailed provided they are controlled by the specified assembly requirements or by separate detail drawings.

4.3 Requirements

An assembly drawing includes, as applicable:

(a) two or more parts, subordinate assemblies, or combination of these items;

(b) a parts list specifying the unique identifier for all items which become a part of the assembly (see 1.7);

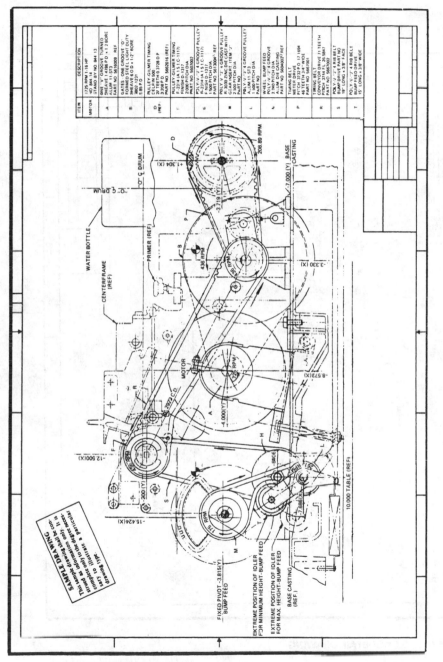

FIG. 4 LAYOUT DRAWING   22

TYPES AND APPLICATIONS OF ENGINEERING DRAWINGS   ASME Y14.24M-1989

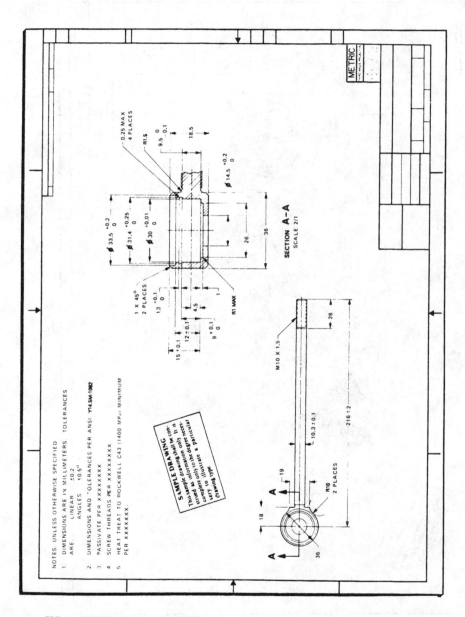

FIG. 5   MONODETAIL DRAWING

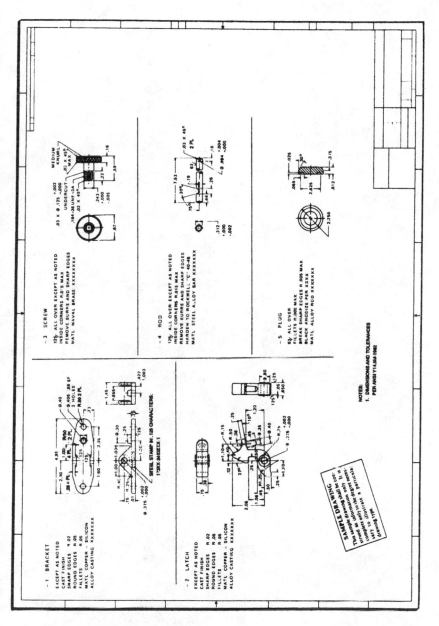

FIG. 7   MULTIDETAIL DRAWING

25

Appendix B

AN AMERICAN NATIONAL STANDARD

Parts Lists, Data Lists, and Index Lists

ASME Y14.34M–1989

[REVISION OF ANSI Y14.34M-1982(R1988)]

 The American Society of
Mechanical Engineers

345 East 47th Street, New York, N.Y. 10017 –

FOREWORD

This Foreword is not part of ASME Y14.34M–1989)

The ASME Y14 Committee, Standards for Engineering Drawing and Related Documentation Practices, authorized the preparation of the original edition of this Standard in September 1972. The Y14 Subcommittee 34, Parts Lists, Data Lists, and Index Lists, was organized and it proceeded to develop the draft on which this Standard is based. The first edition of this Standard was approved by the American National Standards Institute on August 12, 1982, and issued as ANSI Y14.34M–1982.

It is well recognized that individual companies have many detailed requirements for their specific method of operation. Consequently, the minimum requirements set forth in this Standard will provide them flexibility in implementation.

The basis for this Standard is Chapter 600 of DoD-STD-100, Military Standard Engineering Drawing Practices. Every effort has been made to maintain the basic requirements of DoD-STD-100 so as not to invoke a hardship on either the procuring activities or the using companies in the defense community when this Standard is invoked on government contracts.

Following approval by the Y14 Committee and ASME, this Standard was approved as an American National Standard on December 28, 1989.

CONTENTS

ASME Y14.34M-1989

ENGINEERING DRAWING AND RELATED DOCUMENTATION PRACTICES

PARTS LISTS, DATA LISTS, AND INDEX LISTS

1 SCOPE

This Standard establishes the minimum requirements for the preparation and revision of parts lists, data lists, and index lists. In addition, this Standard presents certain options that may be incorporated into parts lists, data lists, and index lists, at the discretion of the design activity.

2 APPLICABLE DOCUMENTS

When the following American National Standards referred to in this Standard are superseded by a revision approved by the American National Standards Institute, Inc., the revision shall apply.

American National Standards
ANSI Y14.1–1980(R1987), Engineering Drawings and Related Documentation Practices, Drawing, Sheet Size and Format
IEEE STD 200–1975 (ANSI Y32.16–1975), Reference Designations for Electrical and Electronic Parts and Equipment

US Government Document
Cataloging Handbook H4/H8, Commercial and Government Entity (CAGE)

3 DEFINITIONS

alpha-numeric arrangement — a group of mixed letter-number designations arranged so that the character furthest to the left in each designation is aligned. All characters in this first portion are arranged alphabetically where possible. Below the alphabetized characters, those with numbers in the first position are arranged consecutively (0-9). Further organization of those characters in the second and succeeding positions is achieved by placing the designations composed only of single characters first, those with a dash second, the alphabetized group third, and finally, the numerically arranged group.

authentication — an indication that the document meets general requirements for preparation and content. Unless established by the procedures of the preparing agency or by contractual requirements, an entry does not indicate approval of the design depicted on the drawings.

Automated Data Processing System (ADPS) — a system used to collect, process, and reproduce data in a selected format through the use of electronic data processing equipment, electrical accounting machine, or other automated equipment

CAGE code — a code assigned by the Government to identify a firm which designs, manufactures or supplies items for Government usage

commercial requirement — a term indicating a requirement when a list is being prepared to meet industry standards and not necessarily to meet government standards

data list (DL) — a tabulation of all engineering drawings, associated lists, specifications, standards, and subordinate data lists pertaining to the item to which the data list applies and essential in-house documents necessary to meet the technical design disclosure requirements except for those in-house documents referenced parenthetically

design activity — an activity having responsiblity for the design of an item. The activity may be a government activity or a contractor, vendor, or other.

find number or item number — a reference number assigned to an item in lieu of the item's identifying number on the field of a drawing and entered as a cross reference to the line of the parts list where the actual item name and identification number are given. Reference designations in accordance with IEEE STD 200-1975 (ANSI Y32.16) may be used as find numbers or item numbers.

flag note — a note whose text is prefixed by a note identification enclosed within a symbol (flag). The

1

note is cross-referenced to a specific area on a drawing, or associated list, by entering the flag at the point of application.

government requirement — a term indicating a requirement when a list is being prepared to meet government specifications

index list (IL) — a tabulation of data lists and subordinate index lists pertaining to the item to which the index applies

item — a nonspecific term used to denote any unit or product including parts, materials, assemblies, equipment, accessories, and attachments

parts list (PL) — a tabulation of all parts and bulk materials (except those materials that support a process) used in the item. Referenced documents may also be tabulated on parts lists. Items listed on subordinate assembly parts lists or specified in a referenced document are not repeated in the using assembly parts list unless it is necessary to limit options permitted by the subordinate document. In-house documents, for in-house usage only, may be referenced parenthetically.

NOTE: Other items previously used to describe a parts list are: list of materials, bill of materials, stock list, and item list.

(a) *integral parts list* — a parts list prepared and revised as part of an engineering drawing

(b) *separate parts list* — a parts list prepared as a document separate from the engineering drawing with which it is associated and one that may be revised independently of the drawing. A single parts list for a product that lists lower assemblies and all components of those assemblies is not addressed by this Standard.

revision authorization — A revision authorization is a document such as a Notice of Revision, Engineering Change Notice, or Revision Directive which describes a revision to a drawing or associated document and is issued by the activity having revision authority.

4 COMMON REQUIREMENTS

4.1 Block and Column Size and Arrangement

The size and arrangement of all blocks and columns shall be determined by the preparing activity according to the method of preparation used.

4.2 Issue/Revision Identification

Each sheet of a list shall show the revision level applicable to that sheet. All sheets of a drawing revised at the same time shall carry the same revision level. Integral parts list sheets shall record revision status in accordance with the system applicable to their parent drawing. When a separate parts list, data list, or index list is to be revised and reissued in its entirety, all sheets shall show the same revision level, including those that have not been revised.

4.2.1 Revision status is indicated by one of the following methods:

(a) On commercial programs, revisions may be identified by sequentially assigned numbers starting with the number one (1) for the first revision. The method used for government programs may also be used.

(b) On government programs, revisions shall be identified by uppercase letters assigned in alphabetic sequence. The first revision to a list shall be assigned the letter "A". The letters "I", "O", "Q", "S", "X", and "Z" shall be omitted. When revisions have exhausted the letter of the alphabet, the revision following "Y" shall be "AA", and the next "AB", the next "AC", etc. Should "AA" to "AY" be exhausted, the next sequence shall be "BA", "BB", etc. Revision letters shall not exceed two characters. The initial release of a list shall be indicated by the entry of a dash (-) in the revision block.

4.3 Sheet Numbering

Multisheet lists shall be numbered consecutively. Numbering shall start with the number one (1) or its decimal equivalent, such as 1.0, 1.00, 1.000, etc. The first sheet shall indicate the total number of sheets. On ADPS prepared sheets the first or last sheet shall indicate the total number of sheets.

4.4 Cover Sheets

Lists may utilize a manually prepared cover sheet as sheet one of the list. The cover sheet may contain required approvals, drawing control information, revision record, and other information.

4.5 Revisions

Existing items or entries may be revised manually or by ADPS.

Appendix **C**

AN AMERICAN NATIONAL STANDARD

Digital Representation for Communication of Product Definition Data

ASME Y14.26M–1989
(REVISION OF ASME/ANSI Y14.26M-1987)

 The American Society of
Mechanical Engineers

345 East 47th Street, New York, N.Y. 10017

Foreword

This standard is based on the work of the Initial Graphics Exchange Specification/Product Data Exchange Specification (IGES/PDES) Organization that is chaired by the National Institute of Standards and Technology (formerly the National Bureau of Standards). Currently, the IGES/PDES Organization consists of over 700 individuals from throughout industry, government, and academia.

The need for a means to exchange product definition data began in the late 1970's with the growth of mini-computer based CAD/CAM (Computer Aided Design (or Drafting) and Computer Aided Manufacturing) systems.

The first version of IGES was completed in January 1980. It was submitted to ASME Y14.26 in May 1980 and was approved as an American National Standard in September 1981. Due to the evolving technology in the CAD/CAM field the standard (and the specification) soon became outdated. The IGES Organization continued to extend and enhance its specification. The second version of this specification was completed in July 1982. For various non-technical reasons it was not submitted for standards action. The third version of the specification was completed in July 1986. It was submitted to ASME Y14.26 and was subsequently approved as an American National Standard in September 1987.

The specification (IGES) continues to evolve in order to keep pace with the need to communicate digital product definition data. The fourth version of this specification was completed in June 1988. It was submitted to ASME Y14.26 in October 1988 and was approved as an American National Standard in November 1989.

This 1989 version of the standard, based on IGES Version 4.0, contains many technical extensions and reflects a desire to expand the capability to communicate a wider range of product definition data developed and used in today's CAD/CAM systems. In addition, refinements to the document improve the syntax, clarity, and precision of the text and figures thus providing the reader with a better guideline for implementation. It should be noted that no technical changes have been made to the specification during the standardization process. However, some administrative and editorial changes were made thus making the standard a more useful reference document.

It is useful for the reader to be aware of some of the major changes to the standard. These changes are briefly described below.

Basic geometry entities have had no format changes from the 1987 version. However, minor changes to their descriptions have been made to clarify existing entities. For example, the weights of B-spline curves and surfaces must now be greater than zero.

Classic drafting entities, including annotation and some geometry entities have also remained as in the 1987 version. Some clarifications and error corrections have been made to the annotation section. New forms have been added to the Ordinate Dimension Entity (Type 222, Form 1), and General Symbol Entity (Type 228, Form 1, 2, and 3). It should also be noted that the new General Symbol forms are to be used to add feature control information to the entity.

A major enhancement is the addition of Constructive Solid Geometry (CSG) to the standard. Specifically, the ways of representing the regularized operations for union, intersection, and difference have been defined. Primitive representations have been established for block, wedge, cylinder, cone, sphere, torus, solid of linear extrusion, solid of revolution, and ellipsoid. It should be noted that the other popular approach to solids, called Boundary Representation (B-rep), is not included as there is yet no generally agreed upon set of definitions.

Several entities that were added to version 4.0 of the specification did not receive adequate testing and were relegated to appendix J. In keeping with the original intent of the specification, we have included appendix J in this standard as a nonmandatory appendix. Three major applications that are affected by the addition of these entities are electrical/electronics, Architecture/Engineering/Construction (AEC), and Finite Element Modeling (FEM). The areas of change are outlined below:

Electrical/electronic applications can be extended to include the ability to attach pre-defined electrical attributes and properties. Nominal values, maximum ratings, propagation delays, and other pertinent data about bipolar transistors and other devices may be defined using the parameters contained in the Attribute Table Entity (Type 322, Electrical Attribute List, ALT=2).

Two major contributions were included for AEC applications. First, the Attribute Table Entities (Type 322 and Type 422) can be used to define attribute data and associated graphic representations that are often necessary for the various AEC applications (e.g., a pattern fill used to denote where concrete is used for construction purposes). Second, Appendix D contains a three-dimensional piping model that shows how existing entities can be used to define piping information.

The Finite Element Modeling (FEM) capability can be expanded to describe FEM results data. Specifically, the Nodal Results Entity (Type 146) can be used for defining temperature and displacement results and the Element Results Entity (Type 148) can be used for defining elemental stress and strain results.

Lastly, it is worth noting that a supplemental index is included in the standard. This index contains a numerical index of the entities which provides quick access in obtaining needed information about a specific entity type.

Contents

CONTENTS

CONTENTS

CONTENTS

CONTENTS

CONTENTS

CONTENTS

CONTENTS

List of Figures

LIST OF FIGURES

LIST OF FIGURES

List of Tables

1.   General

1.1 Purpose

This Standard establishes information structures to be used for the digital representation and communication of product definition data. Use of this Standard permits the compatible exchange of product definition data used by various Computer-Aided Design and Computer-Aided Manufacturing (CAD/CAM) systems.

1.2 Field of Application

This Standard defines a file structure format, a language format, and the representation of geometric, topological, and non-geometric product definition data in these formats. Product definition data represented in these formats will be exchanged through a variety of physical media. The specific features and protocols for the communications media are the subject of other standards. The methodology for representing product definition data in this Standard is extensible and independent of the modeling methods used.

Chapter 1 is general in nature and defines the overall purpose and objectives of this Standard. Chapter 2 defines the communications file structure and format. It explains the function of each of the sections of a file. The geometry data representation in Chapter 3 deals with two- and three-dimensional edge-vertex models, with simple surface representations and Constructive Solid Geometry (CSG) representations. Chapter 4 specifies non-geometric representations, including common drafting practices, data organization methods, and data definition methods.

In Chapters 3 and 4, the product is described in terms of geometric and non-geometric information, with non-geometric information being divided into annotation, definition, and organization. The geometry category consists of elements such as points, curves, surfaces, and solids that model the product. The annotation category consists of those elements which are used to clarify or enhance the geometry, including dimensions, drafting notation, and text. The definition category provides the ability to define specific properties or characteristics of individual or collections of data entities. The organization category identifies groupings of elements from geometric, annotation, or property data which are to be evaluated and manipulated as single items.

1.3 Untested Entities

It is the policy of the IGES/PDES Organization and the ASME Y14.26 Committee to ensure that entities are tested before being included in the Standard. In cases where this testing is not yet complete, the entity is included in Appendix J. A prospective implementor is warned that, despite the fact that Appendix J entities represent the best judgment of the organization, there is a chance that changes will be required before these entities can be added to the Standard. If these entities

1. GENERAL

are judged useful and implementation is attempted, the results of the attempt will be useful to the IGES/PDES Organization. Contact the IGES/PDES Administration Office at the National Institute of Standards and Technology to report problems and successes.

1.4 Concepts of Product Definition

This Standard is concerned with the data required to describe and communicate the essential engineering characteristics of physical objects as manufactured products. Such products are described in terms of their physical shape, dimensions, and information which further describes or explains the product. The processes which generate or utilize the product definition data typically include design, engineering analysis, production planning, fabrication, material handling, assembly, inspection, marketing, and field service.

The requirements for a common data communication format for product definition can be understood in terms of today's CAD/CAM environment. Traditionally, engineering drawings and associated documentation are used to communicate product definition data. Commercial interactive graphics systems, originally developed as aids to producing these two-dimensional drawings, are rapidly developing sophisticated three-dimensional solid modeling. In parallel, extensive research work is being conducted in advanced geometric modeling techniques (*e.g.*, parametric representations and solid primitives) and in CAM applications utilizing product definition data in manufacturing (*e.g.*, NC machining and computer-controlled coordinate measurement). The result is rapid growth of CAD/CAM applications, allowing exchange of product definition data, which usually employ incompatible data representations and formats. In addressing this compatibility problem, this Standard is concerned with needs and capabilities of current and advanced methods of CAD/CAM product definition development.

Product definition data may be categorized by their principal roles in defining a product. An example of such a categorization is presented in Figure 1. This Standard specifies communication formats (information structures) for subsets of the product definition.

1.5 Concepts of the File Structure

A format to allow the exchange of a product definition between CAD/CAM systems must, as a minimum, support the communication of geometric data, annotation, and organization of the data. The file format defined by this Standard treats the product definition as a file of entities. Each entity is represented in an application-independent format, to and from which the native representation of a specific CAD/CAM system can be mapped. The entity representations provided in this Standard include forms common to the CAD/CAM systems currently available and forms which support the system technologies currently emerging.

The fundamental unit of data in the file is the entity. Entities are categorized as geometry and non-geometry. Geometry entities represent the definition of the physical shape and include points, curves, surfaces, solids, and relations which are collections of similarly structured entities. Non-geometry entities typically serve to enrich the model by providing a viewing perspective in which a planar drawing may be composed and by providing annotation and dimensioning appropriate to the drawing. Non-geometry entities further serve to provide specific attributes or characteristics for individual or groups of entities and to provide definitions and instances for groupings of entities. The definitions of these groupings may reside in another file. Typical non-geometry entities for drawing definition, annotation, and dimensioning are the view, drawing, general note, witness line, and leader. Typical non-geometry entities for attributes and groupings are the property and the associativity entities.

2

1.5. CONCEPTS OF THE FILE STRUCTURE

- ADMINISTRATIVE

  Product Identification

  Product Structure

- DESIGN/ANALYSIS

  Idealized Models

- BASIC SHAPE

  Geometric

  Topological

- AUGMENTING PHYSICAL CHARACTERISTICS

  Dimensions and Tolerances

  Intrinsic Properties

- PROCESSING INFORMATION

- PRESENTATIONAL INFORMATION

Figure 1. Categories of Product Definition

A file consists of five or six sections: Flag (in the case of the binary or compressed ASCII form), Start, Global, Directory Entry, Parameter Data, and Terminate. A file may include any number of entities of any type as required to represent the product definition. Each entity occurrence consists of a directory entry and a parameter data entry. The directory entry provides an index and includes descriptive attributes about the data. The parameter data provides the specific entity definition. The directory data are organized in fixed fields and are consistent for all entities to provide simple access to frequently used descriptive data. The parameter data are entity-specific and are variable in length and format. The directory data and parameter data for all entities in the file are organized into separate sections, with pointers providing bi-directional links between the directory entry and parameter data for each entity. The Standard provides for groupings whose definitions will be found in a file other than the one in which they are used.

Each entity defined by the file structure in Chapter 2 has a specific assigned entity type number. While not all are assigned at this time, entity type numbers 0001 through 0599 and 0700 through 5000 are allocated for specific assignment. Entity type numbers 0600 through 0699 and 10000 through 99999 are for implementor-defined (*i.e.*, macro) entities. For user defined entities see Section 1.6.7. The Index of Topics includes an alphabetical listing of entity types.

Some entity types include a form number as an attribute. The form number serves to further define or classify an entity within its specific type.

The entity-set includes a provision for associativities and properties. The Associativity Entity provides a mechanism to establish relationships among entities and to define the meaning of the relationship. The Property Entity allows specific characteristics, such as line widening, to be assigned to an entity or collection of entities. Each entity format includes a structure for an arbitrary

3

1. GENERAL

number of pointers to associativities and properties. The file structure provides for both predefined associativities and properties to be included in the Standard and unique definitions which will be defined by the user.

1.6 Concepts of Information Structures for Geometric Models

The geometric model refers to the entity set defined by Chapters 3 and 4, and comprises an entity-based product definition file. The entity types, as described above are categorized as geometry and non-geometry. In general, the geometry entities are defined independently of one another (surfaces are an exception). Features have been provided to define and compose relationships among entities to enhance the model. The non-geometry entities include structures in which an entity may be defined by a collection of other entities and structures which are independent.

Several entity types which are used to provide relations or definitions are essential to the file structure methodology of this Standard and are described below.

1.6.1 Property Entity. The Property Entity allows non-geometric numeric or textual information to be related to any entity. Any entity occurrence may reference one or more property entity occurrences as required. In addition, a value which is contained in a property may be displayed as text when an additional pointer (See Section 2.2.4.4.2) of the property points to a Text Display Template Entity (Type 432).

Property Entities themselves may exist independently of other entities. In this case, the property is defined to be a property of the level indicated in the level field of the directory entry (DE) of the property. This allows for a general property to apply to all entities of a given level or for the assignment of an applications function to a level. Because the level field in a DE is also allowed to point to a property of levels, properties may be applied to multiple levels.

1.6.2 Associativity Entities. The Associativity Entities are designed for use when several entities must be logically related to one another. Two types of entities are involved here: Associativity Definition and Associativity Instance. The Associativity Definition Entity is used to specify the structure of the logical relationship, and the Associativity Instance Entity is used to specify the information involved in a particular occurrence of the relationship. Some associativities are specifically defined as part of this Standard in Section 4.3.3.3.

1.6.3 View Entity. A drawing or equivalent human-readable representation of the geometric model of a product is a two-dimensional projection of a selected subset of the model, together with non-geometric information such as text. The View Entity and Views Visible forms of associativities control such representations. These provide information for orientation, clipping, line removal, and other characteristics associated with individual views rather than with the model itself.

1.6.4 Drawing Entity. The Drawing Entity allows a set of views to be identified and arranged for human presentation. Note that the View and Drawing Entities contain only the rules and parameters for extracting drawings from the geometric model. The actual product definition is not duplicated in various views, eliminating risk of conflicting or ambiguous information.

1.6.5 Transformation Matrix Entity. The Transformation Matrix Entity allows translation and rotation to be applied as needed to any entity in the construction of the model and to the development of views and drawings of the model.

4

1.6.6 Macro Entities. This Standard includes a Macro Definition Entity for defining new entity types which may then be used in the defining file in the same manner as the intrinsically defined entities. A language for defining these new entity types is specified in Section 4.3.6.

1.6.7 Implementor Defined Entities. This Standard allows implementors to include entities in their files that are not defined in this document but which have specific user meanings. This feature supports the objective of the Standard to act as an archiving format where the receiving system is the same as the sending system. In this way, the implementor is able to archive those data forms which may be unique to a particular system.

From time to time, files with such implementor-defined entities are used with applications which attempt to edit the file. In this situation, processing problems can arise because, without an entity definition, the editor cannot know which parameter values are pointers that have to be updated, and which are simply data values that should not be updated.

To avoid this problem, implementors should use macro definitions and instances of Macro Entities with entity type numbers in the range of 5001 to 9999 inclusively. (See Section 4.3.6 for information on how to use the macro capabilities of the Standard.) This means that for each different implementor-defined entity type, there will be a Macro Definition Entity (Type 306). In order to accomplish the desired result, all that needs to be present in the parameter data for these macro definitions is the first MACRO statement which defines the parameter list, and an ENDM statement to terminate the definition.

1.7 Appendices

As an aid to the implementor/user, a series of nonmandatory appendices is included. Appendix A gives three part file examples. Appendix B describes an electrical/electronic product representation, and Appendix C, a plant flowsheet representation. Appendix D provides a three-dimensional piping model example while Appendix E gives explanation of spline representation and approaches for conversion techniques. Appendix F discusses the numerical stability of conic arcs. Appendix G provides mappings between color spaces. Appendix H provides a set of FORTRAN utilities to convert physical file structures in the ASCII Form from the regular ASCII Format to the Compressed ASCII Format and back. Appendix I itemizes entities from previous versions which have been made obsolete by this version. Appendix J includes new entities which have not received sufficient implementation testing for inclusion in the main body of the Standard.

Appendix A.   Part File Examples

This appendix is not part of ASME Y14.26M-1989, *Digital Representation for Communication of Product Definition Data*, and is included for information purposes only.

This appendix contains three sample parts encoded in the ASCII Form. These files are included to provide a guide to the usage of IGES and this Standard and, as such, do not represent all design application uses. The files are a two-dimensional application using structure entities, a two-dimensional drawing of a mechanical part with dimensioning, and a three-dimensional part with two-dimensional drawing views defined.

Example file 1 is an integrated circuit (IC) cell. The IC application was selected because of the predominance of two-dimensional geometry used in electrical designs. The geometry used in the cell in Figure A1 consists of simple closed area, linear path entities and line widening property. The structure entities are nested subfigures using a Network Subfigure Definition Entity and Array Subfigure Instance Entities. A Connect Point Entity is included to identify the signal port. The geometry is on five different levels, each representing a process mask. The entity label field of each Directory Entry record contains (optional) text included to describe the entity's use. The entities in this file would be typical of those used in an IC application to transfer either cell libraries or a complete design between design systems. The file of a design prepared for pattern generation, with subfigures resolved and the geometry fractured, would use the Flash Entity exclusively. The cell file was adapted from a cell library in [HON80] with kind permission from the author.

Example file 2 is a two-dimensional drawing of a mechanical part containing geometry entities and annotation entities typically found on engineering drawings. Included as geometry are points, lines, circular arcs and conics. For annotation, the file includes linear dimensions, angular dimensions, radius dimensions, ordinate dimensions, a general label and general notes. Figure A2 shows the mechanical part, which was used during one of the early public demonstrations of intersystem data exchange.

Example file 3 is included to show the use of View Entities and Drawing Entities in conjunction with a three-dimensional part model to convey a drawing to the receiving system. Figure A3 shows the example drawing. In this way, model geometry and viewing parameters are logically separate. A three-dimensional model, as well as the drawing, is received enabling additional views to be created if necessary, and changes to the part model are reflected in all views.

APPENDIX A. ELECTRICAL PART EXAMPLE

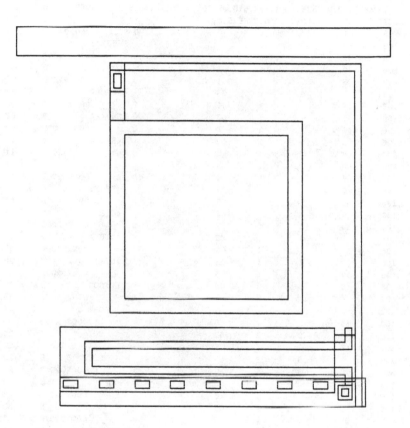

Figure A1. Electrical Part Example

ASME Y14.26M-1989

APPENDIX A. ELECTRICAL PART EXAMPLE

Example 1 Electrical Part

```
INTEGRATED CIRCUIT SEMICUSTOM CELL (ONE PART OF A LIBRARY FILE)        S     1
USED IN APPENDIX A OF IGES VERSION 3.0 AND MODIFIED FOR VERSION 4.0     S     2
1H,,1H;,10H5MICRONLIB,5HPADIN,9HEXAMPLE 1,4HIGES,16,38,06,38,13,        G     1
10HIC.LIBRARY,1.0,9,2HUM,1,,13H880616.090000,0.01,265.0,9HBILL LOYE,    G     2
33HELECTRICAL APPLICATIONS COMMITTEE,5,0;                               G     3
```

308	01		1	0		0	00020201D	01
308	0		1			SUBFIG1	D	02
106	02		1	1		0	00020200D	03
106	0	4	1	63		VDDPORT	D	04
106	03		1	1		0	00020200D	05
106	0	4	1	63		GNDPORT	D	06
106	04		1	1		0	00020200D	07
106	0	4	1	63		BONDPAD	D	08
106	05		1	7		0	00020200D	09
106	0	5	1	63		GLASSBOX	D	10
320	06		1	0		0	00010201D	11
320	0		1			CELLFIG	D	12
408	07		1	0		0	00030200D	13
408	0		1			INST1	D	14
106	08		1	3		0	00020200D	15
106	0	3	2	63		ACTBOX	D	16
106	10		1	3		0	00020200D	17
106	0	3	1	63		ACTBOX	D	18
106	11		1	3		0	00020200D	19
106	0	3	1	11		ACTSTG	D	20
106	12		1	3		0	00020200D	21
106	0	3	2	63		ACTBOX	D	22
132	14		1	3		0	00020400D	23
132	0		1			SIGPORT	D	24
106	15		1	6		0	00020200D	25
106	0	8	2	63		CUT	D	26
106	17		1	6		0	00020200D	27
106	0	8	2	63		CUT	D	28
308	19		1	0		0	00020201D	29
308	0		1			SUBFIG2	D	30
106	20		1	6		0	00030200D	31
106	0	8	1	63		CUTDEF	D	32
412	21		1	0		0	00030201D	33
412	0		1			CUTARR	D	34
106	22		1	2		0	00020200D	35
106	0	2	2	11		GATESTG	D	36
106	24		1	2		0	00020200D	37
106	0	2	2	63		GATEBOX	D	38
106	26		1	1		0	00020200D	39
106	0	4	1	63		GATEBOX	D	40
406	27		1	0		0	00010200D	41
406	0		1	5		LINWIDTH	D	42

353

APPENDIX A. ELECTRICAL PART EXAMPLE

```
308,0,6HPADBLK,4,03,05,07,09;                                      01P    01
106,1,5,0.,0.,0.,265.,0.,265.,-20.,0.,-20.,0.,0.;                  03P    02
106,1,5,0.,30.,-245.,245.,-245.,245.,-265.,30.,-265.,30.,-245.;    05P    03
106,1,5,0.,65.,-65.,200.,-65.,200.,-200.,65.,-200.,65.,-65.;       07P    04
106,1,5,0.,75.,-75.,190.,-75.,190.,-190.,75.,-190.,75.,-75.;       09P    05
320,1,5HPADIN,11,13,15,17,19,21,25,27,33,35,37,39,2,,,1,23;        11P    06
408,01,0.,0.,0.;                                                   13P    07
106,1,5,0.,30.,-210.,222.5,-210.,222.5,-255.,30.,-255.,30.,        15P    08
-210.;                                                             15P    09
106,1,5,0.,65.,-25.,75.,-25.,75.,-45.,65.,-45.,65.,-25.;           17P    10
106,1,3,0.,77.5,-27.5,240.,-27.5,240.,-262.5,0,1,41;               19P    11
106,1,5,0.,222.5,-215.,237.5,-215.,237.5,-247.5,222.5,-247.5,      21P    12
222.5,-215.;                                                       21P    13
132,240.,-265.,0.,,2,1,,,,,01,2,1,11;                              23P    14
106,1,5,0.,67.5,-32.5,72.5,-32.5,72.5,-42.5,67.5,-42.5,67.5,       25P    15
-32.5;                                                             25P    16
106,1,5,0.,227.5,-252.5,232.5,-252.5,232.5,-257.5,227.5,-257.5,    27P    17
227.5,-252.5;                                                      27P    18
308,0,7HCONTACT,1,31;                                              29P    19
106,1,5,0.,-5.,2.5,5.,2.5,5.,-2.5,-5.,-2.5,-5.,2.5;                31P    20
412,29,1.0,37.5,-250.,0.,8,1,25.,0.,0.,0;                          33P    21
106,1,6,0.,232.5,-212.5,232.5,-222.5,50.,-222.5,50.,-240.,         35P    22
232.5,-240.,232.5,-247.5,0,1,41;                                   35P    23
106,1,5,0.,225.,-250.,235.,-250.,235.,-260.,225.,-260.,225.,       37P    24
-250.;                                                             37P    25
106,1,5,0.,65.,-30.,75.,-30.,75.,-65.,65.,-65.,65.,-30.;           39P    26
406,5,5.0,1,1,0,0;                                                 41P    27
S     2G     3D     42P     27                                     T      1
```

ASME Y14.26M-1989

APPENDIX A. MECHANICAL PART EXAMPLE

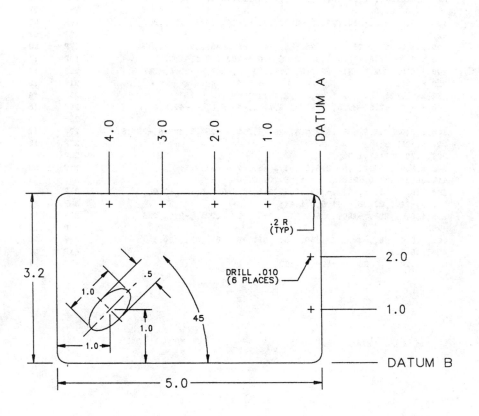

Figure A2. Mechanical Part Example

APPENDIX A. MECHANICAL PART EXAMPLE

Example 2 Mechanical Part Example

```
PLATE.001    SAMPLE MECHANICAL PART WITH ANNOTATION                    S    1
   USED AT AUTOFACT - OCTOBER 1982 AND IN VERSION 3.0 APPENDIX A       S    2
      AUTHOR: Dave Briggs, Boeing                                       S    3
ENTITY CONTENT:  POINT, LINE, ARC and CONIC                             S    4
                 LINEAR, ANGULAR, RADIUS, POINT and ORDINATE DIMENSION  S    5
                 GENERAL NOTE, GENERAL LABEL                            S    6
CHANGE LOG:                                                             S    7
,,8HPANEL123,10HPANEL.IGES,4HEX 2,4HIGES,16,38,7,38,14,8HPANEL123,1.0,  G    1
1,4HINCH,1,0.028,13H880516.144004,0.0005,100.0,9HD. BRIGGS,6HBOEING,4,0;G 2
```

124	1	1	1	0	0	0	0	0D	1
124	0	0	2	0				0D	2
212	3	1	1	5	0	0	0	10100D	3
212	0	0	2	0				0D	4
214	5	1	1	5	0	0	0	10100D	5
214	0	0	1	2				0D	6
210	6	1	1	5	0	0	0	100D	7
210	0	0	1	0				0D	8
110	7	1	1	1	0	0	0	0D	9
110	0	0	1	0				0D	10
110	8	1	1	1	0	0	0	0D	11
110	0	0	1	0				0D	12
110	9	1	1	1	0	0	0	0D	13
110	0	0	1	0				0D	14
110	10	1	1	1	0	0	0	0D	15
110	0	0	1	0				0D	16
100	11	1	1	1	0	0	0	0D	17
100	0	0	1	0				0D	18
100	12	1	1	1	0	0	0	0D	19
100	0	0	1	0				0D	20
100	13	1	1	1	0	0	0	0D	21
100	0	0	1	0				0D	22
100	14	1	1	1	0	0	0	0D	23
100	0	0	1	0				0D	24
116	15	1	1	2	0	0	0	0D	25
116	0	0	1	0				0D	26
116	16	1	1	2	0	0	0	0D	27
116	0	0	1	0				0D	28
116	17	1	1	2	0	0	0	0D	29
116	0	0	1	0				0D	30
116	18	1	1	2	0	0	0	0D	31
116	0	0	1	0				0D	32
104	19	1	1	3	0	1	0	0D	33
104	0	0	2	1				0D	34
116	21	1	1	2	0	0	0	0D	35
116	0	0	1	0				0D	36
116	22	1	1	2	0	0	0	0D	37
116	0	0	1	0				0D	38

ASME Y14.26M-1989

APPENDIX A. DRAWING AND VIEW EXAMPLE

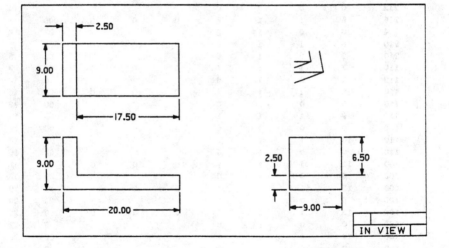

Figure A3. Drawing and View Example

ASME Y14.26M-1989

APPENDIX A. DRAWING AND VIEW EXAMPLE

Example 3 Drawing and View Example

```
Test file of model with VIEW (410) and DRAWING (404) entities          S    1
                                                                       S    2
This file demonstrates annotation attached to the VIEWS,               S    3
i.e., the dimension entities are flagged as INDEPENDENT,               S    4
and their DE field 6 points to a VIEW entity. The coordinates          S    5
of the dimensions are in MODEL space, and they have a                  S    6
transformation matrix which is the inverse of the VIEW matrix.         S    7
                                                                       S    8
This file was submitted by:                                            S    9
                                                                       S   10
        Dennette A. Harrod, Jr.                                        S   11
        Computervision Corporation                                     S   12
        14 Crosby Drive / Bldg. 5-1                                    S   13
        Bedford, MA   01730                                            S   14
        617-275-1800 ext.5172                                          S   15
                                                                       S   16
1H,,1H;,8HVIEWDWG2,12HVIEWDWG2.IGS,13H<unspecified>,13H<unspecified>,32,G    1
38,6,38,15,8HVIEWDWG2,1..,1,2HIN,8,1..,13H870930.151125,1.E-06,71.,    G    2
13H<unspecified>,13H<unspecified>,4,0;                                 G    3
```

406	1	1	0	0	0	0	0	10300D	1
406	0	0	1	15				0D	2
124	2	1	0	0	0	0	0	10300D	3
124	0	0	2	0				0D	4
108	4	1	0	0	0	0	0	10201D	5
108	0	0	2	1				0D	6
108	6	1	0	0	0	0	0	10201D	7
108	0	0	2	1				0D	8
108	8	1	0	0	0	0	0	10201D	9
108	0	0	2	1				0D	10
108	10	1	0	0	0	0	0	10201D	11
108	0	0	2	1				0D	12
410	12	1	0	0	0	3	0	201D	13
410	0	0	1	0				0D	14
406	13	1	0	0	0	0	0	10300D	15
406	0	0	1	15				0D	16
124	14	1	0	0	0	0	0	10300D	17
124	0	0	1	0				0D	18
108	15	1	0	0	0	0	0	10201D	19
108	0	0	1	1				0D	20
108	16	1	0	0	0	0	0	10201D	21
108	0	0	1	1				0D	22
108	17	1	0	0	0	0	0	10201D	23
108	0	0	1	1				0D	24
108	18	1	0	0	0	0	0	10201D	25
108	0	0	1	1				0D	26
410	19	1	0	0	0	17	0	201D	27
410	0	0	1	0				0D	28

363

Appendix B.   Electrical/Electronic Product Representation

This appendix is not part of ASME Y14.26M-1989, *Digital Representation for Communication of Product Definition Data*, and is included for information purposes only.

B.1   Introduction

B.1.1   Purpose. The purpose of this appendix is to provide implementors and users with a roadmap to the representation of electrical/electronic product designs using this Standard. The topics of discussion will include (but are not limited to) design, engineering, manufacturing, testing, and inspection.

B.1.2   Assumptions. The reader should have a basic understanding of electrical/electronic product design using CAD/CAM and Computer-Aided Engineering (CAE) tools, including (but not limited to) connectivity, component descriptions, placement and routing, and the manufacturing interface. Each topic will be discussed in general, but these discussions are not intended to provide a tutorial on the subject.

B.2   Connectivity

B.2.1   General. Forming a connection between two or more items requires the ability to represent the following:

1. the exact location of each connection point,

2. the signal formed and its identification (if any), and

3. the physical connection between the items (if any).

The term "connect node" will refer to a database entity which represents the exact location of connection. The term "link" will refer to the representation of the signal formed, and "signal name" will refer to the signal identifier. The term "join" will refer to the database entity or entities which represent the physical connection between the items.

Each item to be connected requires a connect node to represent each possible connection point of the item. A signal may be formed between any such items by a link which references the connect nodes to be connected. This creates an associativity between the connect nodes, and thus the connected items. The signal name may be used to uniquely identify the particular signal formed. The join may be used to provide a graphical representation of the signal. In electrical applications, the join will most often be represented by a line (schematic) or a widened line (printed wiring board).

APPENDIX B.2 CONNECTIVITY

In electrical applications, the items to be connected are typically electrical components (*i.e.*, resistor, 16-pin DIP, *etc.*). Most often, these components are represented by subfigures which are defined once, then referenced (instanced) in the database for each occurrence of the component. Each pin of the component is a potential connection point in a signal; thus each subfigure has a connect node defined for each pin. When such a subfigure is instanced, its connect nodes must also be instanced. This allows each connect node to participate in the unique signal to which it belongs. An instanced connect node, when added to a signal, is different from its definition which is not a member of any signal.

These subfigures, representing electrical components, often contain text describing the component and its pins. In some cases (*e.g.*, part number), this text is fixed and unchanging. In other cases (*e.g.*, reference designator), the text may be variable, and thus may not be filled in until the subfigure is instanced. This text (sometimes called a "text node"), like the connect node, is instanced along with its parent subfigure. In some cases, a connect node and a text node are related (*e.g.*, the connect node represents a component pin and the text node represents the pin number).

B.2.2 Representation. The connect node is represented by the Connect Point Entity. The text node is represented by the Text Display Template Entity. The Flow Associativity Entity represents a signal and contains the link, signal name, and pointers to the join entities. The Network Subfigure Definition and Instance represent electrical components which participate in signals. A number of Property Entities will also be used.

B.2.2.1 Network Subfigure Construction. A component is constructed using the Network Subfigure Definition Entity. The entities representing the component geometry are referenced in the same manner as the Subfigure Definition Entity. In addition, a separate set of pointers to defining Connect Point Entities is provided. These Connect Point Entities define the locations and characteristics of the component's pins. Properties, for example the Part Name Property, may be attached to the Network Subfigure Definition Entity.

B.2.2.2 Connect Points. A component pin (or surface mounted device pad, IC I/0 port, lead frame, schematic symbol lead, *etc.*) is represented by the Connect Point Entity. The Connect Point Entity is used in both logical and physical product designs. The exact location in model space is specified, along with several characteristic flags (connection type, function type, I/0 direction). There is a pointer to the parent Network Subfigure Entity (definition or instance), which provides a association needed for signal processing. An additional Subfigure Instance pointer is provided for Connect Point display. This allows a graphical symbol to be displayed, representing the Connect Point. The pin number is provided in the Function Connect Point Identifier field, along with a pointer to a Text Display Template for pin number display. A pin function name is provided in the Connect Point Function Name Field, along with a pointer to a Text Display Template for its display.

B.2.2.3 Signal Construction. A signal, representing one set of electrically common Connect Points, is constructed using the Flow Associativity Entity. It contains pointers to other associated Flow Associativity Entities, the Connect Point Entities participating in the signal (this is the Link), and the Join entities representing the geometry of the signal (either logical or physical). Also contained is a list of signal names which may be used to identify the signal, along with a set of pointers to Text Display Template Entities which may be used to display the first signal name in a number of locations. Two characteristic flags determine the signal type (logical or physical), and the function type (fluid flow or electrical signal).

ASME Y14.26M-1989

APPENDIX B.2 CONNECTIVITY

A signal is represented by one Flow Associativity Entity pointing to a set of electrically common Connect Points. This is the link. The Join entities represent the physical display geometry of the signal. For a schematic (logical), a line without width is typically used. For a printed board (physical), a line with the Line Widening Property is typically used. A number of signal names may be associated with the signal. Multiple displays of the first, or primary, name are possible.

The components participating in a signal are represented by the Network Subfigure Instance Entity. Note that the Connect Point Entities "belonging" to a component are instanced along with the subfigure. This is necessary to allow a subfigure to participate in a number of different signals, while retaining unique component/pin identification. Each component is usually identified by a reference designator. This is supplied by the Primary Reference Designator Field of the Network Subfigure. Any alternate reference designators may be designated with the Reference Designator Property, attached through the normal property pointer mechanism (see Section 2.2.4.4.2).

B.2.2.4 Information Display. Throughout the above discussion, references to the Text Display Template Entity have been made. This entity allows text, embedded in an entity, to be displayed without the redundant specification of the text string. There are two reasons for this feature. First, it eliminates a possible source of error by allowing the information to be specified in only one place. Second, it reduces the file size overhead. This entity exists in two forms, absolute and incremental. The absolute form provides an exact location for display of the information, as in the display of a reference designator. The incremental form provides an offset to be applied to the parent entity's location to provide the exact location for display of information like pin numbers. When a direct pointer for information display is provided, the base location is readily determined from the parent entity's location such as when displaying a pin number. In the case of property value display, the base location must be determined by "remembering" the location of the property entity's parent entity. This would occur when displaying the Part Name. Also in this case, the pointer to the Text Display Template Entity is located in the additional pointers section of the property entity parameters.

B.2.2.5 Additional Considerations. The situation is exactly the same for both logical and physical representations. The only differences arise in the Subfigure and Join entities used. In fact, a file may contain representations for both the schematic and the board. The Flow Associativity Entity contains a type flag to indicate the connection type (logical or physical). In this case, one Flow Associativity would represent the logical connection and a second would represent the physical connection. The two associativities would be related by the pointers provided in the Flow Associativity.

B.2.2.6 Figures. The following figures illustrate certain aspects of the above discussions. Figure B1 illustrates the basic entity relationships. Figure B2 and Table B1 illustrate the usage of the Text Display Template. Figure B3 illustrates an actual implementation. Figure B4 shows an example of logical and physical signals and their relationships in the same file.

                                                    ASME Y14.26M-1989

APPENDIX B.3 ELECTRICAL ENTITY DESCRIPTIONS

Table B1. Text Display Template Values for Sample Schematic

TEMPLATE	TLEFT	TRIGHT	TTOP	TBOT	
WIDTH	.10	.10	.10	.10	
HEIGHT	.13	.13	.13	.13	
SLANT ANGLE	(default)	(default)	(default)	(default)	
ROT'N. ANGLE	0.	0.	0.	0.	
MIRROR FLAG	0	0	0	0	
VRT./HORIZ.	0	0	0	0	
DE FORM NO.	1	1	1	1	
X (DX)	-.09	-.03	+.03	+.03	
Y (DY)	+.03	+.03	-.15	+.1	
Z (DZ)	0.	0.	0.	0.	
U1	P1–P8	P9–P16	—	—	Pins using
U2	P1–P8	P9–P16	—	—	the text
U3	P1–P4	P9–P12	P13–P16	P5 P8	templates

B.3   Electrical Entity Descriptions

The following entities (in entity number order) are the subset of entities which have particular meanings when used for electrical product data.

100 Circular Arc Entity

The electrical use of this entity is in the geometric representation of component parts and their symbolic representations. In such usage, it is generally part of a subfigure. It may be used as a join entity. Its use may be defined by a Level Function Property or DE Level Pield.

102 Composite Curve Entity

The electrical use of this entity is in the geometric representation of component parts and their symbolic representations. In such usage, it is generally part of a subfigure. It may be used as a join entity. Its use may be defined by a Level Function Property or DE Level Field.

106 Copious Data Entity

Forms 11, 12, and 13 of this entity may be used to implement the electrical join (*i.e.*, schematic or wiring diagram circuit connection lines). Any of these forms may point to a Line Widening Property. Examples of this entity property combination are circuit paths on a printed circuit (PC) board or integrated circuit (IC) metalization, or as a bus in a schematic. Form 63, Simple Closed Area, may be used to define an auto-router restriction area or a PC (or IC) defined area having special attributes.

ASME Y14.26M-1989

APPENDIX B.3 ELECTRICAL ENTITY DESCRIPTIONS

108 Plane Entity

Certain layers of PC design are designated as "ground", "power", or "heat sink", and as such are large conductive areas. These layers, as well as larger curved or rounded conductive areas on other layers, are best defined by the Plane Entity. Note that the form number indicates whether the bounded region is positive or void (*i.e.*, copper clad area or cutout).

110 Line Entity

The Line Entity has several important uses in the electrical application. It may be used to construct component outlines, and to represent both logical and physical connections (as a join entity). As a physical join entity, the Line Widening Property will most often be attached, giving the width of metalization to be etched on the board. As a logical join entity, the line will most typically be used without the Line Widening Property.

116 Point Entity

The Point Entity is used to locate features that do not participate in electrical connectivity, for example, a mounting hole.

124 Transformation Matrix Entity

A Transformation Matrix Entity may be used to rotate subfigures to other than normal (top up) positions. Generally, rotations are about the Z axis for PC and IC constructs, but may be about any axis for three-dimensional cabling files.

125 Flash Entity

The Flash Entity may be used to represent a repetitive artwork feature which is usually produced by photo-optical means. Examples include PC pads, targets, clearances, and IC features.

132 Connect Point Entity

The Connect Point Entity is used to represent a point of connection. A subfigure defining an electrical component typically uses the Connect Point Entity to represent a pin of the logical or physical component or symbol. A Connect Point may also be used in a "stand-alone" mode, representing a via hole, for example.

212 General Note Entity

A General Note Entity is used to display constant text. Design notes would require a General Note Entity.

ASME Y14.26M-1989

APPENDIX B.3 ELECTRICAL ENTITY DESCRIPTIONS

302 Associativity Definition Entity

When the originating system provides a relationship not included among the predefined associativities, this entity is required. Possible uses are to relate subfigures to entities in other databases (*i.e.*, circuit analysis or text requirements) or for back-annotation purposes.

312 Text Display Template Entity
Form 0: absolute; Form 1: incremental

The Text Display Template may be used to display text which may be unique in each instance of the defined entity (*i.e.*, a pin number).

320 Network Subfigure Definition Entity

For printed circuit and cable usage, a subfigure usually represents a component and its required PC constructs. This entity is normally a library physical or logical figure in the originating system.

402 Associativity Instance Entity

This entity relates other entities within a file to provide a "set" with a common meaning. Those associativities which are predefined are identified by form numbers (*i.e.*, Form 18: Flow). The user defined associativities are defined by Entity 302 and its form number.

406 Property Entity

The use of a property to indicate the meaning or purpose of a geometric entity applies to electrical constructs as well as general graphics. A Connect Point Entity may point to the Drilled Hole Property. A Plane Entity or Simple Closed Area Entity may point to the Region Restriction Property. Any property, however, may point to the Text Display Template with the text string specified in the property. In this case, the Text Display Template locates the text display. This entity is an open-ended list allowing for expansion to address future needs such as simulation, testing, inspection applications, and extensions into electrical/electronic systems hierarchical design.

412 Rectangular Array Subfigure Instance Entity
414 Circular Array Subfigure Instance Entity

These entities may be used to instance multiple Network Subfigure Instances, but must not be used for instancing Connect Points.

420 Network Subfigure Instance Entity

This entity allows an electrical component to be instanced in a number of unique locations. Note that "owned" Connect Point Entities must be instanced with this entity.

ASME Y14.26M-1989

APPENDIX B.3 ELECTRICAL ENTITY DESCRIPTIONS

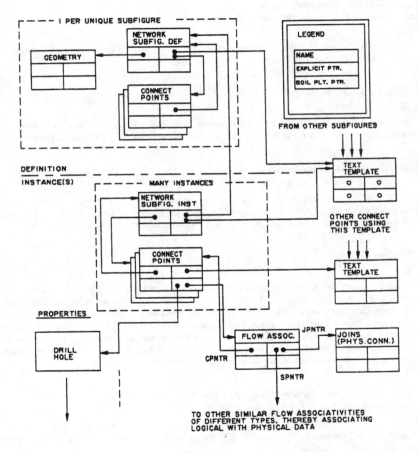

Figure B1. General Pointer and Entity Model

APPENDIX B.3 ELECTRICAL ENTITY DESCRIPTIONS

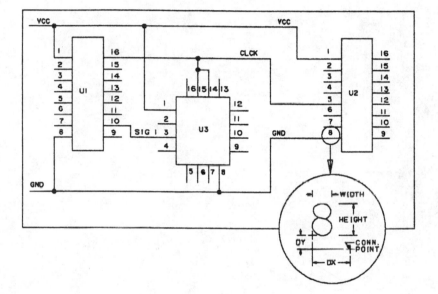

Figure B2. Sample Schematic

ASME Y14.26M-1989

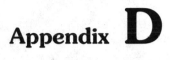

Appendix **D**

AMERICAN NATIONAL STANDARD
ENGINEERING DRAWINGS AND RELATED
DOCUMENTATION PRACTICES

Dimensioning and Tolerancing

ANSI Y14.5M - 1982

SECRETARIAT

THE AMERICAN SOCIETY OF MECHANICAL ENGINEERS
SOCIETY OF AUTOMOTIVE ENGINEERS

PUBLISHED BY

THE AMERICAN SOCIETY OF MECHANICAL ENGINEERS

United Engineering Center   345 East 47th Street   New York, N.Y. 10017

FOREWORD

(This Foreword is not a part of American National Standard, Engineering Drawings and
Related Documentation Practices, Dimensioning and Tolerancing, ANSI Y14.5M-1982.)

This issue is a revision of American National Standard Y14.5-1973, Dimensioning and Tolerancing. Changes reflected herein are intended to improve and update U.S. dimensioning and tolerancing practices and to seek a closer alignment with corresponding practices adopted internationally.

Several considerations were initially involved in determining the extent of this revision. Foremost among them were: comments deferred from the previous issue, metrication, symbology, and international standardization. Others evolved as the work progressed.

Members of the Y14.5 Subcommittee have participated in a number of international meetings on dimensioning and tolerancing since 1973. The outcome of these meetings has influenced the direction of change in certain areas and broadened the base upon which this Standard was developed.

Planning meetings were held in London in 1973, 1974, and 1975 to prepare for the next, and what proved to be the last, American-British-Canadian-Australian (ABCA) Conference on Unification of Engineering Standards (UES). The conference convened in Ottawa in 1977. Agreements were reached on unifying certain practices to narrow existing differences between the four national standards of the major English-speaking countries.

The ABCA/UES activity, founded in the latter years of World War II, was officially discontinued in 1980 due to the increased growth and tempo of international standardization work. Nevertheless, the liaison established between ABCA countries on subjects covered by this Standard is expected to continue, although on a less formal basis.

At the seventh meeting of ISO/TC 10 at Scheveningen in 1972, the International Organization for Standardization (ISO) Committee on Technical Drawings (TC 10) outlined a program of work for the years to follow. A U.S. delegation attended this meeting and submitted ten papers on subjects proposed for consideration in the program of work for Subcommittee 5 (SC 5), Dimensioning and Tolerancing.

Subsequent meetings of ISO/TC 10/SC 5 and its associated Working Groups were held in Zurich (1973), Oslo (1974), London (1975), New York (1976), West Berlin (1976), Ottawa (1977), Copenhagen (1978), Zurich (1979), and Cologne (1980). U.S. participation in these activities has contributed greatly toward the dissemination of useful ideas and the advancement of international standardization.

Several techniques introduced in the Y14.5 Standard have gained ISO acceptance. They are the projected tolerance zone, the three-plane datum concept, total runout tolerance, multiple datums, and datum targets. Consideration of other U.S. proposals is anticipated as unresolved subjects are addressed by the Working Groups.

During the development of this revision, joint meetings were held with the U.S. Technical Advisory Group for ISO/TC 10/SC 5, consisting of those Y14.5 members authorized to serve as U.S. delegates in ISO activities. Emerging international practices were discussed and carefully evaluated by the Subcommittee as items for adoption. Those considered to be practical and technically appropriate are incorporated herein. Some were disapproved and shall likely remain points of difference.

A long-prevailing example is the exception this Standard has taken to the ISO symbols and methods for indicating a datum feature. Practicality of the U.S. symbol and method was reaffirmed by the Subcommittee and the practice continued without change.

Additionally, a conflict in principle has arisen with regard to the interpretation of the limits of size. As of publication time of this Standard, ISO is proceeding with the adoption of a "principle of independency," which absolves size limits from the control of form of an individual feature. This is contrary to the Taylor* principle, wherein a boundary of perfect form at MMC is prescribed to control variations in form as well as size of individual features.

If the principle of independency is adopted by ISO, size limits on drawings prepared in accordance with ISO standards would be verified by two-point measurements only. Consequently, if it were necessary to control the form of a feature by means of its size limits, an "envelope" principle (boundary of perfect form at MMC) would be invoked by indicating the symbol Ⓔ following the size dimension.

The Y14.5 Standard continues to subscribe to the Taylor principle, which is currently recognized and widely accepted in the U.S. and elsewhere. Special indication of its application is not required on U.S. drawings since it is a fundamental precept used in product, tool, and gage design.

The technical expertise provided by the Y14.5 Subcommittee is attributable to a broad cross section of U.S. industry, the Department of Defense (DoD), educational institutions, and national laboratories. Particular recognition is given to those organizations who have sponsored the travel and participation of individual members of the Subcommittee in national and international standardization activities.

The following is a summary of the principal changes and improvements incorporated in this issue of American National Standard Y14.5.

- By definition, distinction is made between a feature and a feature of size: see 1.3.7 and 1.3.8.

- Definitions are given for *datum feature* and *geometric tolerance*: see 1.3.9 and 1.3.19.

- Implied 90° angles and unspecified 90° basic angles are clarified: see 1.4 (i) and (j).

- SI (metric) linear units are featured throughout the text and in illustrations: see 1.5. By request of ASME, the publisher of this Standard, the spelling of millimetre was changed to millimeter.

- Millimeter dimensioning practices are now covered under general dimensioning: see 1.6.1.

- The international method for indicating a limited length or area of a surface (which is to receive a particular consideration) is introduced: see 1.7.3.

- The use of parentheses to identify a reference dimension (or reference data) is standardized: see 1.7.6.

- The diameter symbol always precedes each diameter value: see 1.8.1.

- The radius symbol always precedes each radius value: see 1.8.2.

- The international method for specifying repetitive features and dimensions is introduced: see 1.9.5.

---

*William Taylor, an Englishman, whose patent in 1905 featured full form "GO" gages. See ISO 1938, Part 1.

- Methods are established for introducing tolerance requirements expressed in the international system of limits and fits: see 2.2.1.

- Millimeter tolerancing practices are now covered under general tolerancing: see 2.3.1.

- A symbol is introduced for indicating the origin for dimensional limits in certain applications: see 2.6.1.

- Explanation of the control of geometric form prescribed by limits of size is restructured: see 2.7.1.

- For positional tolerancing, the practice of always specifying whether RFS, MMC, or LMC applies to an individual tolerance, datum reference, or both is standardized: see 2.8. The alternate practice, identified as Rule #2a in the 1973 issue, is discontinued.

- The effect of LMC applied to a positional tolerance and datum reference is explained: see 2.8.3.

- International practices employing unique symbols are introduced for specifying the taper of a cone and the slope of a flat taper as ratios: see 2.13 and 2.14.

- Geometric characteristics categorized as "form" tolerances in the 1973 issue are now classified as "form," "profile," or "orientation" tolerances: see Fig. 68.

- The total runout symbol previously used is replaced by a new symbol that is adopted internationally: see Fig. 68.

- Use of the symmetry symbol is discontinued. The position symbol is used where symmetry applies.

- The rearrangement of entries within the datum target symbol is in accordance with international practice: see 3.3.3.

- Modifying symbols used to indicate "at least material condition," "spherical diameter," "spherical radius," and "arc" are introduced: see 3.3.5, 3.3.7, and 3.3.9.

- New symbols are introduced to replace certain terms frequently specified on drawings. These terms are *counterbore* or *spotface*, *countersink*, *depth*, and *square*: see 3.3.10, 3.3.11, 3.3.12, and 3.3.13.

- The international sequence of entries within a feature control frame (formerly called feature control symbol) is standardized: see 3.4.2.

- A symbol used to indicate "all around" in profile tolerance applications is introduced: see 3.4.2.3.

- Criteria are provided for establishing datums from datum features: see 4.4.

- The explanation of datum targets is expanded: see 4.5.

- The leader used to direct a datum target symbol to the target location is replaced by a radial line: see 4.5.1.

- A positional tolerancing example is included which symmetrically locates a pattern of features relative to the center planes of datum features of size: see Fig. 121.

- Zero positional tolerancing at MMC, formerly explained in Appendix B of the 1973 issue, is incorporated in this Standard: see 5.3.3.

- The LMC principle is introduced for positional tolerancing applications: see 5.3.5.

v

- The composite positional tolerancing method used for specifying pattern-locating and feature-relating requirements is standardized: see 5.4. The combination of positional and plus and minus tolerancing methods is discontinued.

- Bidirectional positional tolerancing of circular features is introduced: see 5.9.

- A zero positional tolerance at MMC is featured for the control of coaxiality and symmetry of related features within their limits of size: see 5.11.1.3 and 5.12.1.2.

- An example of a positional tolerance applied to a spherical feature is included: see 5.13.

- Recognition is given to the practice of applying profile tolerances "all over" to parts such as castings: see 6.5.2.

- Additional examples are provided for profile tolerances applied to plane and conical surfaces: see 6.5.7 and 6.5.8.

- The section on dual dimensioning (Section 5-7 in the 1973 issue) has been deleted.

- The scope of Appendix A is broadened by providing information on dimensioning for computer-aided design and computer-aided manufacturing.

- Information on the advantages of positional tolerancing (Appendix B in the 1973 issue) has been deleted.

- Appendix B is essentially the same as Appendix C in the 1973 issue, except for the values given in millimeters and an example of formula usage where requirements are expressed by symbols for limits and fits.

- In Appendix C, symbol proportions are given as a factor of letter height instead of frame height as in Appendix D in the 1973 issue.

- Information on former practices once featured in the 1966 and 1973 issues of this Standard is provided in the new Appendix D.

A draft of this issue was released for Y14, industry, and DoD review in December 1980. All comments received were discussed and resolved by the Y14.5 Subcommittee. Authors of comments have been individually notified of the results. Following approval by the Y14 Standards Committee and the Co-Secretariats, this issue was approved by the American National Standards Institute on December 16, 1982.

Suggestions for improvement of this Standard will be welcomed. They should be sent to the American National Standards Institute, 1430 Broadway, New York, New York 10018.

CONTENTS

ANSI Y14.5M-1982

AMERICAN NATIONAL STANDARD
ENGINEERING DRAWINGS AND RELATED DOCUMENTATION PRACTICES

DIMENSIONING AND TOLERANCING

1 Scope, Definitions, and General Dimensioning

1.1 GENERAL

This Standard covers dimensioning, tolerancing, and related practices for use on engineering drawings and in related documents. Uniform practices for stating and interpreting these requirements are established herein. Practices unique to architectural and civil engineering, as well as welding symbology, are not included.

1.1.1 Units. The International System of Units (SI) is featured in this Standard because SI units are expected to supersede United States (U.S.) customary units specified on engineering drawings. It should be understood that customary units could equally well have been used without prejudice to the principles established.

1.1.2 Reference to This Standard. Where drawings are based on this Standard, this fact shall be noted on the drawings or in a document referenced on the drawings. References to this Standard shall state "ANSI Y14.5M-1982."

1.1.3 Figures. The figures in this Standard are intended only as illustrations to aid the user in understanding the principles and methods of dimensioning and tolerancing described in the text. The absence of a figure illustrating the desired application is neither reason to assume inapplicability nor basis for drawing rejection. In some instances figures show added detail for emphasis, in other instances figures are incomplete by intent. Numerical values of dimensions and tolerances are illustrative only.

1.1.4 Notes. Notes herein in capital letters are intended to appear on finished drawings. Notes in lower case letters are explanatory only and are not intended to appear on drawings.

1.1.5 Reference to Gaging. This document is not intended as a gaging standard. Any reference to gaging is included for explanatory purposes only.

1.1.6 Symbols. Adoption of the symbols indicating dimensional requirements as illustrated in Fig. C2 of Appendix C does not preclude the use of equivalent terms or abbreviations where symbology is considered inappropriate.

1.2 REFERENCES

When the following American National Standards referred to in this Standard are superseded by a revision approved by the American National Standards Institute, Inc., the revision shall apply.

American National Standards
ANSI B4.2-1978, Preferred Metric Limits and Fits
ANSI B46.1-1978, Surface Texture
ANSI B87.1-1965, Decimal Inch
ANSI B89.3.1-1972, Measurement of Out-of-Roundness
ANSI B94.6-1981, Knurling
ANSI B94.11M-1979, Twist Drills
ANSI Y14.1-1980, Drawing Sheet Size and Format
ANSI Y14.2M-1979, Line Conventions and Lettering
ANSI Y14.6-1978, Screw Thread Representation
ANSI Y14.9-1958, Forgings
ANSI Y14.36-1978, Surface Texture Symbols
ANSI Z210.1-1976, Metric Practice

1.3 DEFINITIONS

The following terms are defined as their use applies in this Standard.

1.3.1 Dimension. A numerical value expressed in appropriate units of measure and indicated on a drawing and in other documents along with lines, symbols, and notes to define the size or geometric characteristic, or both, of a part or part feature.

1.3.2 Basic Dimension. A numerical value used to describe the theoretically exact size, profile, orientation, or location of a feature or datum target. It is the basis from which permissible variations are established by tolerances on other dimensions, in notes, or in feature control frames (see Fig. 78).

1.3.3 True Position. The theoretically exact location of a feature established by basic dimensions.

1.3.4 Reference Dimension. A dimension, usually without tolerance, used for information purposes only. It is considered auxiliary information and does not govern production or inspection operations. A reference dimension is a repeat of a dimension or is derived from other values shown on the drawing or on related drawings.

1.3.5 Datum. A theoretically exact point, axis, or plane derived from the true geometric counterpart of a specified datum feature. A datum is the origin from which the location or geometric characteristics of features of a part are established.

1.3.6 Datum Target. A specified point, line, or area on a part used to establish a datum.

1.3.7 Feature. The general term applied to a physical portion of a part, such as a surface, hole, or slot.

1.3.8 Feature of Size. One cylindrical or spherical surface, or a set of two plane parallel surfaces, each of which is associated with a size dimension.

1.3.9 Datum Feature. An actual feature of a part that is used to establish a datum.

1.3.10 Actual Size. The measured size.

1.3.11 Limits of Size. The specified maximum and minimum sizes.

1.3.12 Maximum Material Condition (MMC). The condition in which a feature of size contains the maximum amount of material within the stated limits of size—for example, minimum hole diameter, maximum shaft diameter.

1.3.13 Least Material Condition (LMC). The condition in which a feature of size contains the least amount of material within the stated limits of size—for example, maximum hole diameter, minimum shaft diameter.

1.3.14 Regardless of Feature Size (RFS). The term used to indicate that a geometric tolerance or datum reference applies at any increment of size of the feature within its size tolerance.

1.3.15 Virtual Condition. The boundary generated by the collective effects of the specified MMC limit of size of a feature and any applicable geometric tolerances.

1.3.16 Tolerance. The total amount by which a specific dimension is permitted to vary. The tolerance is the difference between the maximum and minimum limits.

1.3.17 Unilateral Tolerance. A tolerance in which variation is permitted in one direction from the specified dimension.

1.3.18 Bilateral Tolerance. A tolerance in which variation is permitted in both directions from the specified dimension.

1.3.19 Geometric Tolerance. The general term applied to the category of tolerances used to control form, profile, orientation, location, and runout.

1.3.20 Full Indicator Movement (FIM). The total movement of an indicator when appropriately applied to a surface to measure its variations.

1.4 FUNDAMENTAL RULES

Dimensioning and tolerancing shall clearly define engineering intent and shall conform to the following.

(*a*) Each dimension shall have a tolerance, except for those dimensions specifically identified as reference, maximum, minimum, or stock (commercial stock size). The tolerance may be applied directly to the dimension (or indirectly in the case of basic dimensions), indicated by a general note, or located in a supplementary block of the drawing format (see ANSI Y14.1).

(*b*) Dimensions for size, form, and location of features shall be complete to the extent that there is full understanding of the characteristics of each feature. Neither scaling (measuring the size of a feature directly from an engineering drawing) nor assumption of a distance or size is permitted.

NOTE: Undimensioned drawings — for example, loft, printed wiring, templates, master layouts, tooling layout — prepared on stable material are excluded, provided the necessary control dimensions are specified.

(*c*) Each necessary dimension of an end product shall be shown. No more dimensions than those necessary for complete definition shall be given. The use of reference dimensions on a drawing should be minimized.

(*d*) Dimensions shall be selected and arranged to suit the function and mating relationship of a part and shall not be subject to more than one interpretation.

(*e*) The drawing should define a part without specifying manufacturing methods. Thus, only the diameter of a hole is given without indicating whether it is to be drilled, reamed, punched, or made by any other operation. However, in those instances where manufacturing, processing, quality assurance, or environmental information is essential to the definition of engineering requirements, it shall be specified on the drawing or in a document referenced on the drawing.

(*f*) It is permissible to identify as nonmandatory certain processing dimensions that provide for finish allowance, shrink allowance, and other requirements, provided the final dimensions are given on the drawing. Nonmandatory processing dimensions shall be identified by an appropriate note, such as NONMANDATORY (MFG DATA).

(*g*) Dimensions should be arranged to provide required information for optimum readability. Dimensions should be shown in true profile views and refer to visible outlines.

(*h*) Wires, cables, sheets, rods, and other materials manufactured to gage or code numbers shall be specified by linear dimensions indicating the diameter or thickness. Gage or code numbers may be shown in parentheses following the dimension.

(*i*) A 90° angle is implied where center lines and lines depicting features are shown on a drawing at right angles and no angle is specified (see 2.1.1.2).

(*j*) A 90° BASIC angle applies where center lines of features in a pattern or surfaces shown at right angles on the drawing are located or defined by basic dimensions and no angle is specified.

(*k*) Unless otherwise specified, all dimensions are applicable at 20°C (68°F). Compensation may be made for measurements made at other temperatures.

1.5 UNITS OF MEASUREMENT

For uniformity, all dimensions in this Standard are given in SI units. However, the unit of measurement selected should be in accordance with the policy of the user.

1.5.1 SI (Metric) Linear Units. The commonly used SI linear unit used on engineering drawings is the millimeter.

1.5.2 U.S. Customary Linear Units. The commonly used U.S. customary linear unit used on engineering drawings is the decimal inch.

1.5.3 Identification of Linear Units. On drawings where all dimensions are either in millimeters or inches, individual identification of linear units is not required. However, the drawing shall contain a note stating UNLESS OTHERWISE SPECIFIED, ALL DIMENSIONS ARE IN MILLIMETERS (or IN INCHES, as applicable).

1.5.3.1 Where some inch dimensions are shown on a millimeter-dimensioned drawing, the abbreviation IN. shall follow the inch values. Where some millimeter dimensions are shown on an inch-dimensioned drawing, the symbol mm shall follow the millimeter values.

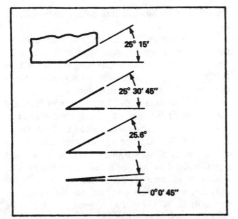

FIG. 1 ANGULAR UNITS

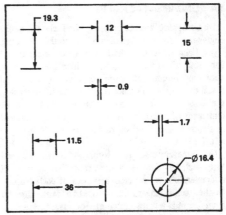

FIG. 2 MILLIMETER DIMENSIONS

1.5.4 Angular Units. Angular dimensions are expressed in either degrees and decimal parts of a degree or in degrees, minutes, and seconds. These latter dimensions are expressed by symbols: for degrees °, for minutes ', and for seconds ". Where degrees are indicated alone, the numerical value shall be followed by the symbol °. Where only minutes or seconds are specified, the number of minutes or seconds shall be preceded by 0° or 0°0', as applicable. See Fig. 1.

1.6 TYPES OF DIMENSIONING

Decimal dimensioning shall be used on drawings except where certain commercial commodities are identified by standardized nominal designations such as pipe and lumber sizes.

1.6.1 Millimeter Dimensioning. The following shall be observed when specifying millimeter dimensions on drawings.

(a) Where the dimension is less than one millimeter, a zero precedes the decimal point. See Fig. 2.

(b) Where the dimension is a whole number, neither the decimal point nor a zero is shown. See Fig. 2.

(c) Where the dimension exceeds a whole number by a decimal fraction of one millimeter, the last digit to the right of the decimal point is not followed by a zero. See Fig. 2.

NOTE: This practice differs for tolerances expressed bilaterally or as limits [see 2.3.1(b) and (c)].

(d) Neither commas nor spaces shall be used to separate digits into groups in specifying millimeter dimensions on drawings.

1.6.2 Decimal Inch Dimensioning. The decimal inch system is explained in ANSI B87.1. The following shall be observed when specifying decimal inch dimensions on drawings.

(a) A zero is not used before the decimal point for values less than one inch.

(b) A dimension is expressed to the same number of decimal places as its tolerance. Zeros are added to the right of the decimal point where necessary. See Fig. 3 and 2.3.2.

1.6.3 Decimal Points. Decimal points must be uniform, dense, and large enough to be clearly visible and meet the reproduction requirements of ANSI Y14.2M. Decimal points are placed in line with the bottom of the associated digits.

1.6.4 Conversion and Rounding of Linear Units. For information on conversion and rounding of U.S. customary linear units, see ANSI Z210.1.

1.7 APPLICATION OF DIMENSIONS

Dimensions are applied by means of dimension lines, extension lines, chain lines, or a leader from a dimension, note, or specification directed to the appropriate feature. See Fig. 4. General notes are used to convey

AMERICAN NATIONAL STANDARD
DIMENSIONING AND TOLERANCING

ANSI Y14.5M-1982

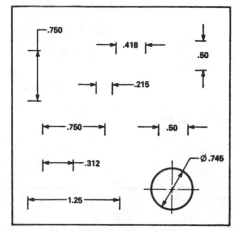

FIG. 3   DECIMAL INCH DIMENSIONS

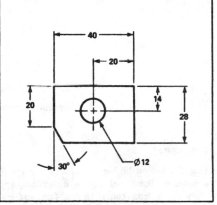

FIG. 4   APPLICATION OF DIMENSIONS

additional information. For further information on dimension lines, extension lines, chain lines, and leaders, see ANSI Y14.2M.

1.7.1 Dimension Lines. A dimension line, with its arrowheads, shows the direction and extent of a dimension. Numerals indicate the number of units of a measurement. Preferably, dimension lines should be broken for insertion of numerals as shown in Fig. 4. Where horizontal dimension lines are not broken, numerals are placed above and parallel to the dimension lines.

1.7.1.1 Dimension lines shall be aligned if practicable and grouped for uniform appearance. See Fig. 5.

1.7.1.2 Dimension lines are drawn parallel to the direction of measurement. The space between the first dimension line and the part outline should be not less than 10 mm; the space between succeeding parallel dimension lines should be not less than 6 mm. See Fig. 6.

NOTE: These spacings are intended as guides only. If the drawing meets the reproduction requirements of the accepted industry or military reproduction specification, nonconformance to these spacing requirements is not a basis for rejection of the drawing.

Where there are several parallel dimension lines, the numerals should be staggered for easier reading. See Fig. 7.

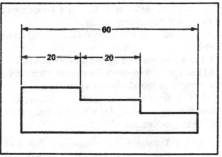

FIG. 5   GROUPING OF DIMENSIONS

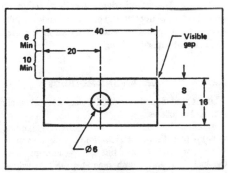

FIG. 6   SPACING OF DIMENSIONS

2 General Tolerancing and Related Principles

2.1 GENERAL

This Section establishes practices for expressing tolerances on linear and angular dimensions, applicability for material condition modifiers, and interpretations governing limits and tolerances.

2.1.1 Application. Tolerances may be expressed as follows:

(a) as direct limits or as tolerance values applied directly to a dimension (see 2.2);

(b) as a geometric tolerance, as described in Sections 5 and 6;

(c) in a note referring to specific dimensions;

(d) as specified in other documents referenced on the drawing for specific features or processes;

(e) in a general tolerance block referring to all dimensions on a drawing for which tolerances are not otherwise specified; see ANSI Y14.1.

2.1.1.1 Tolerances on dimensions that locate features of size may be applied directly to the locating dimensions or specified by the positional tolerancing method described in Section 5.

2.1.1.2 Unless otherwise specified, where a general tolerance note on the drawing includes angular tolerances, it applies to features shown at specified angles and at implied 90° angles.

2.2 DIRECT TOLERANCING METHODS

Limits and directly applied tolerance values are specified as follows.

(a) *Limit Dimensioning.* The high limit (maximum value) is placed above the low limit (minimum value). When expressed in a single line, the low limit precedes the high limit and a dash separates the two values. See Fig. 56.

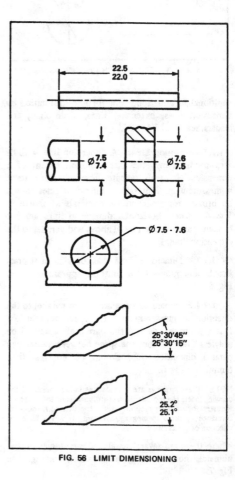

FIG. 56 LIMIT DIMENSIONING

3 Symbology

3.1 GENERAL

This Section establishes the symbols for specifying geometric characteristics and other dimensional requirements on engineering drawings. Symbols should be of sufficient clarity to meet the legibility and reproducibility requirements of ANSI Y14.2M. Symbols are to be used only as described herein.

3.2 USE OF NOTES TO SUPPLEMENT SYMBOLS

Situations may arise where the desired geometric requirement cannot be completely conveyed by symbology. In such cases, a note may be used to describe the requirement, either separately or supplementing a geometric tolerance. See Figs. 135 and 206.

3.3 SYMBOL CONSTRUCTION

Information related to the construction, form, and proportion of individual symbols described herein is contained in Appendix C.

3.3.1 Geometric Characteristic Symbols. The symbols denoting geometric characteristics are shown in Fig. 68.

3.3.2 Datum Feature Symbol. The datum feature symbol consists of a frame containing the datum identifying letter preceded and followed by a dash. See Fig. 69. The symbol frame is related to the datum feature by one of the methods prescribed in 3.5.

3.3.2.1 Letters of the alphabet (except I, O, and Q) are used as datum identifying letters. Each datum feature requiring identification shall be assigned a different letter. When datum features requiring identification on a drawing are so numerous as to exhaust the single alpha series, the double alpha series shall be used – AA through AZ, BA through BZ, etc.

3.3.2.2 Where the same datum feature symbol is repeated to identify that same feature in other locations on a drawing, it need not be identified as reference.

3.3.3 Datum Target Symbol. The datum target symbol is a circle divided horizontally into two halves. See Fig. 70. The lower half contains a letter identifying the associated datum, followed by the target number assigned sequentially starting with 1 for each datum. See Fig. 109. Where the datum target is an area, the area size may be entered in the upper half of the symbol; otherwise, the upper half is left blank. A radial line attached to the symbol is directed to a target point (indicated by an "X"), target line, or target area, as applicable. See 4.5.1.

3.3.4 Basic Dimension Symbol. The symbol used to identify a basic dimension is shown in Fig. 71.

3.3.5 Material Condition Symbols. The symbols used to indicate "at maximum material condition," "regardless of feature size," and "at least material condition" are shown in Fig. 72. The use of these symbols in local and general notes is prohibited.

3.3.6 Projected Tolerance Zone Symbol. The symbol used to indicate a projected tolerance zone is shown in Fig. 72. The use of this symbol in local and general notes is prohibited.

3.3.7 Diameter and Radius Symbols. The symbols used to indicate diameter, spherical diameter, radius, and spherical radius are shown in Fig. 72. These symbols precede the value of a dimension or tolerance given as a diameter or radius, as applicable.

3.3.8 Reference Symbol. A reference dimension (or reference data) is identified by enclosing the dimension (or data) within parentheses. See Fig. 72.

ANSI Y14.5M–1982

4 Datum Referencing

4.1 GENERAL

This Section establishes the principle of datum referencing used to relate features of a part to an appropriate datum or datum reference frame. It contains the criteria for selecting, designating, and using features of a part as the basis for dimensional definition. A datum indicates the origin of a dimensional relationship between a toleranced feature and a designated feature or features on a part. The designated feature serves as a datum feature, whereas its true geometric counterpart establishes the datum.

4.1.1 Application. As measurements cannot be made from a true geometric counterpart which is theoretical, a datum is assumed to exist in and be simulated by the associated processing equipment. For example, machine tables and surface plates, though not true planes, are of such quality that they are used to simulate the datums from which measurements are taken and dimensions verified. When magnified, flat surfaces of manufactured parts are seen to have irregularities; contact is made with a datum plane at a number of surface extremities or high points.

4.1.2 Datum Reference Frame. Sufficient datum features, those most important to the design of a part, are chosen to position the part in relation to a set of three mutually perpendicular planes, jointly called a datum reference frame. This reference frame exists in theory only and not on the part. Therefore, it is necessary to establish a method for simulating the theoretical reference frame from the actual features of the part. This simulation is accomplished by positioning the part on appropriate datum features to adequately relate the part to the reference frame and to restrict motion of the part in relation to it. See Fig. 87.

4.1.2.1 These planes are simulated in a mutually perpendicular relationship to provide direction as well as the origin for related dimensions and measurements. Thus, when the part is positioned on the datum reference frame (by physical contact between each datum feature and its counterpart in the associated processing equipment), dimensions related to the datum reference frame by a feature control frame or note are thereby mutually perpendicular. This theoretical reference frame constitutes the three-plane dimensioning system used for datum referencing.

4.1.2.2 In some cases, a single datum reference frame will suffice. In others, additional datum reference frames may be necessary where physical separation or the functional relationship of features require that datum reference frames be applied at specific locations on the part. In such cases, each feature control frame must contain the datum feature references that are applicable. Any difference in the order of precedence or in the material condition of any datums referenced in multiple feature control frames requires different datum simulation methods and, consequently, establishes a different datum reference frame. See 4.4.9.

4.2 DATUM FEATURES

A datum feature is selected on the basis of its geometric relationship to the toleranced feature and the requirements of the design. To ensure proper part interface and assembly, corresponding features of mating parts are also selected as a datum feature where practicable. Datum features must be readily discernible on the part. Therefore, in the case of symmetrical parts or parts with identical features, physical identification of the datum feature on the

5 Tolerances of Location

5.1 GENERAL

This Section establishes the principles of tolerances of location; included are position, concentricity, and symmetry used to control the following relationships:

(*a*) center distance between such features as holes, slots, bosses, and tabs;

(*b*) location of features [such as in (a) above] as a group, from datum features such as plane and cylindrical surfaces;

(*c*) coaxiality or symmetry of features;

(*d*) features with center distances equally disposed about a datum axis or plane.

5.2 POSITIONAL TOLERANCING

A positional tolerance defines a zone within which the center, axis, or center plane of a feature of size is permitted to vary from true (theoretically exact) position. Basic dimensions establish the true position from specified datum features and between interrelated features. A positional tolerance is indicated by the position symbol, a tolerance, and appropriate datum references placed in a feature control frame.

5.2.1 Method. The following paragraphs describe methods used in expressing positional tolerances.

5.2.1.1 The location of each feature (hole, slot, stud, etc.) is given by basic dimensions. Many drawings are based on a schedule of general tolerances, usually provided near the drawing title block. See ANSI Y14.1. Dimensions locating true position must be excluded from the general tolerance in one of the following ways:

(*a*) applying the basic dimension symbol to each of the basic dimensions [see Fig. 118, parts (a) and (b)];

(*b*) specifying on the drawing (or in a document referenced on the drawing) the general note: UNTOLERANCED DIMENSIONS LOCATING TRUE POSITION ARE BASIC [see Fig. 118, part (c)].

5.2.1.2 A feature control frame is added to the note used to specify the size and number of features. See Figs. 119 through 121. These figures show different types of feature pattern dimensioning.

5.2.1.3 It is necessary to identify features on a part to establish datums for dimensions locating true positions. For example, in Fig. 119, if datum references had been omitted, it would not be clear whether the inside diameter or the outside diameter was the intended datum feature for the dimensions locating true positions. The intended datum features are identified with datum feature symbols and the applicable datum references are included in the feature control frame. For information on specifying datums in an order of precedence, see 4.3.

5.2.2 Application to Base Line and Chain Dimensioning. True position dimensioning can be applied as base line dimensioning or as chain dimensioning. For positional tolerancing, unlike plus and minus tolerancing as shown in Fig. 59, basic dimensions are used to establish the true position of features. Assuming identical positional tolerances are specified, the resultant tolerance between any two holes will be the same for chain dimensioning as for base line dimensioning. This also applies to angular dimensions, whether base line or chain type.

5.3 FUNDAMENTAL EXPLANATION OF POSITIONAL TOLERANCING

The following is a general explanation of positional tolerancing.

6 Tolerances of Form, Profile, Orientation, and Runout

6.1 GENERAL

This Section establishes the principles and methods of dimensioning and tolerancing to control form, profile, orientation, and runout of various geometrical shapes and free state variations.

6.2 FORM AND ORIENTATION CONTROL

Form tolerances control straightness, flatness, circularity, and cylindricity. Orientation tolerances control angularity, parallelism, and perpendicularity. A profile tolerance may control form, orientation, and size, depending on how it is applied. Since, to a certain degree, the limits of size control form and parallelism, and tolerances of location control orientation, the extent of this control should be considered before specifying form and orientation tolerances. See 2.7, Fig. 61, and Fig. 123.

6.3 SPECIFYING FORM AND ORIENTATION TOLERANCES

Form and orientation tolerances critical to function and interchangeability are specified where the tolerances of size and location do not provide sufficient control. A tolerance of form or orientation may be specified where no tolerance of size is given–for example, the control of flatness after assembly of the parts.

6.3.1 Form and Orientation Tolerance Zones. A form or orientation tolerance specifies a zone within which the considered feature, its line elements, its axis, or its center plane must be contained.

6.3.1.1 Where the tolerance value represents the diameter of a cylindrical zone, it is preceded by the diameter symbol. In all other cases, the tolerance

value represents a total linear distance between two geometric boundaries and no symbol is required.

6.3.1.2 Certain designs require control over a limited area or length of the surface, rather than control of the total surface. In these instances, the area, or length, and its location are indicated by a heavy chain line drawn adjacent to the surface with appropriate dimensioning. Where so indicated, the specified tolerance applies within these limits instead of to the total surface.

6.4 FORM TOLERANCES

Form tolerances are applicable to single (individual) features or elements of single features; therefore, form tolerances are not related to datums. The following subparagraphs cover the particulars of the form tolerances: straightness, flatness, circularity, and cylindricity.

6.4.1 Straightness Tolerance. Straightness is a condition where an element of a surface or an axis is a straight line. A straightness tolerance specifies a tolerance zone within which the considered element or axis must lie. A straightness tolerance is applied in the view where the elements to be controlled are represented by a straight line.

6.4.1.1 Figure 170 shows an example of a cylindrical feature where all circular elements of the surface are to be within the specified size tolerance. Each longitudinal element of the surface must lie between two parallel lines separated by the amount of the prescribed straightness tolerance and in a plane common with the nominal axis of the feature. The feature control frame is attached to a leader directed to the surface or extension line of the surface but not to the size dimension. The straightness tolerance must be less than the size tolerance. Since the limits

APPENDIX A

DIMENSIONING FOR COMPUTER-AIDED DESIGN AND COMPUTER-AIDED MANUFACTURING MODE

(This Appendix is not a part of American National Standard, Engineering Drawings and Related Documentation Practices, Dimensioning and Tolerancing, ANSI Y14.5M-1982.)

A1 GENERAL

Industry acceptance of Computer-Aided Design (CAD) and Computer-Aided Manufacturing (CAM) systems for use in component design and fabrication is rapidly accelerating. Collectively, these highly sophisticated systems can be used to describe the desired part as a geometric model, interactively interject manufacturing data, and deliver this information to a designated machine tool for execution of the finished part. Although computer-aided systems continue to require dimensions and tolerances for part definition, in many cases the dimensioning is accomplished by means of algorithms which emulate manual dimensioning practices. In view of the changing state-of-the-art, it is important that the designer understand where certain practices can be employed for expressing dimensional requirements most effectively. The purpose of this Appendix is to contribute to that understanding by first iterating the standard coordinate system and then providing guidelines applicable to the CAD/CAM (data base) mode as well as the manual (conventional drawing) mode. This information will assist the designer in developing dimensioning and tolerancing practices common to these modes.

A2 COORDINATE SYSTEM

The coordinate system is the same for both the geometric model created by CAD and the graphic definition found on conventional drawings. It is the standard system of rectangular or Cartesian coordinates wherein a point is located by its distance from each of two or three mutually perpendicular intersecting planes. Two-dimensional coordinates (in X and Y directions) locate points on a plane, while three-dimensional coordinates (in X, Y, and Z directions) locate points in space. Once a geometric model is defined, it is the basis for interactive programming of commands for the machine tool to execute the required relative motion between cutting tool and workpiece. For CAM usage, dimensional coordinates translate into point locations relative to coordinate axes since linear and rotary motion is involved.

A3 REFERENCE PLANES

For CAD, three mutually perpendicular planes are established from which a geometric model of the desired part can be constructed. These planes normally coincide with the exterior outline of parts having surfaces at right angles. Where the part is cylindrical, two of these planes intersect along the axis of the cylinder and the third is perpendicular to it. When viewed from above, as in the top view of Fig. A1, these planes are oriented in accordance with the following.

(*a*) The first plane lies in the plane of projection. It is the plane from which coordinate distances are specified in the Z direction.

(*b*) The second plane is horizontal and perpendicular to the first. It is the plane from which coordinate distances are specified in the Y direction.

(*c*) The third plane is vertical and perpendicular to the other two. It is the plane from which coordinate distances are specified in the X direction.

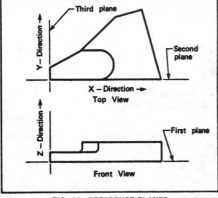

FIG. A1 REFERENCE PLANES

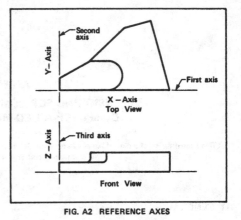

FIG. A2 REFERENCE AXES

A4 REFERENCE AXES

For CAM, three mutually perpendicular axes are established along which linear and rotary motions occur in the machine tool used for producing the desired part. Generally, these axes are designated as the basic coordinate axes of the equipment. Additional (secondary) axes may also be designated, depending on machine capability and part configuration. The basic axes, when viewed from above, as in the top view in Fig. A2, are oriented in accordance with the following.

(*a*) The first axis is horizontal in the plane of projection. It is the X axis of motion.

(*b*) The second axis is vertical in the plane of projection and perpendicular to the X axis. It is the Y axis of motion.

(*c*) The third axis is perpendicular to the plane of projection and perpendicular to the X and Y axes. It is the Z axis of motion.

A5 MATHEMATICAL QUADRANTS

The intersection of the X axis and Y axis forms quadrants described in Fig. A3. These axes are normally aligned or coincident with appropriate surfaces or features on the desired part. When programming commands for the machine tool, the workpiece should be positioned in a quadrant in such a way that a maximum of positive values will result. For example, if the workpiece is positioned in

Quadrant I, positive values will result. If the workpiece is positioned in two or more quadrants, positive and negative values will result, and the potential for error is greater. This precaution is generally not necessary when programming on the computer, but helpful when programming without computer assistance. The considerations described above also apply to quadrants formed by intersections of the X–Z and Y–Z axes.

A6 DIMENSIONING AND TOLERANCING

Recommended guidelines for dimensioning and tolerancing practices for use in defining parts for the CAD/CAM mode are as follows.

(*a*) Major features of a part are used to establish the basic coordinate system for initial part definition. These features may or may not be subsequently identified as datum features.

(*b*) For final part definition, any number of sub-coordinate systems may be used to locate and orient features of a part. These systems, however, must be geometrically related to the basic coordinate system of the given part.

(*c*) Define part features in relation to three mutually perpendicular reference planes. Establish these planes along features which parallel the axes and motions of CAM equipment, wherever possible.

(*d*) The assignment of datum features is based primarily on the functional requirements of the part.

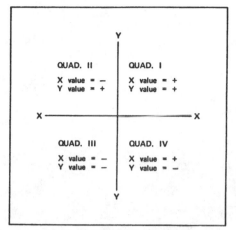

FIG. A3 MATHEMATICAL QUADRANTS

(*e*) Dimension the part so that its geometric shape is completely defined and mathematically precise.

(*f*) Regular geometric profiles such as ellipses, parabolas, hyperbolas, etc., may be defined on the drawing by mathematical formulas. CAM equipment can be programmed to generate these profiles by linear interpolation, that is, a series of short straight lines whose end points are spaced close enough to approximate the desired profile within the specified profile tolerance.

(*g*) A part surface whose profile is defined on the drawing by a mathematical formula should not be coordinately dimensioned unless specific dimensions are required for inspection or identified as reference information.

(*h*) For arbitrary profiles, the drawing should specify appropriate points on the profile by coordinate dimensions, or provide a table of coordinates. When determining the number of points needed to define the profile, keep in mind that the tighter the tolerance or the smaller the radius of curvature, the closer together the points should be. Terms such as "blend smoothly" and "faired curve" are not specified.

(*i*) Profiles may also be defined by other coordinate systems, such as polar, spherical, or cylindrical, as applicable. However, it is desirable to use the same coordinate system on a given drawing.

(*j*) Any change in profile (points of inflection or tangency) should be clearly defined, with prime consideration given to design intent. Precise continuity of the profile is necessary for CAD.

(*k*) A circular pattern of holes may be defined by polar coordinate dimensions. Location and orientation of the pattern must be clearly shown.

(*l*) Where possible, express angular dimensions in degrees and decimal parts of a degree.

(*m*) Limit dimensioning should be avoided except where limits are standardized — for example, preferred "limits and fits."

(*n*) Where plus and minus tolerancing is used, the tolerance should be bilateral and not unilateral. Equal plus and minus values are preferred. Positional tolerancing is recommended for locating features of size.

(*o*) Geometric tolerances are specified in all cases where the control of specific geometric characteristics of part features is required. Where applicable, identifying datum features on the drawing and referencing them in an order of precedence will clearly indicate their usage for CAM.

(*p*) Avoid profile tolerances applied unilaterally along the true profile. Profile tolerances equally disposed bilaterally along the true profile are recommended. Include no fewer than four defined points along the profile.

(*q*) Tolerances should be specified on the basis of actual design requirements. The accuracy capability of CAM equipment is not a basis for specifying more restrictive tolerances than are functionally required.

A7 INCORPORATING DIMENSIONAL CHANGES

Dimensional changes of small magnitude seldom require a change to the graphic definition on conventional drawings. Either the graphic definition remains within acceptable drawing accuracy, or the revised dimensional values are underlined, indicating dimensions "not to scale." For CAD applications, dimensional changes of any magnitude must be made to the data base which requires graphic reiteration. This must be done to ensure mathematical accuracy of changed values and to maintain the integrity of the CAD/CAM data base.

Index

About the Author

Currently program manager for Martin Marietta Energy
Systems in Oak Ridge, Tennessee, and previously Senior
Mechanical Engineer for the Naval Sea Logistics Center,
Kathryn A. Ingle develops and implements new technologies
for commercial applications. Ingle has developed and
implemented reverse engineering pilot programs for the U.S.
Navy, and participates in a variety of American Society of
Mechanical Engineers (ASME) activities on a national level,
including chairing ASME forums and a technical division.